JAVA PROGRAMMING AND WEBSITE DESIGN

Lalit Kishore Arora
M.Tech (C.S.), MCA, B.Sc (Comp. Sc.)
Asstt. Professor, MCA Department
Ajay Kumar Garg Engineering College,
Ghaziabad

S.K. KATARIA & SONS®
Publisher of Engineering Books
4885/109, Prakash Mahal, Dr. Subhash Bhargav Lane,
Opposite Delhi Medical Association, Daryaganj, New Delhi–110002 (INDIA)
Phone: +91-11-43551243; *Mobile:* +91-9871775858, 7678425209
e-mail: katariabook@gmail.com, katariabooks@yahoo.com
Website: www.skkatariaandsons.com
(e-books, subscriptions and bulk buying also available on our website)

Published by:

S.K. Kataria & Sons®
4885/109, Prakash Mahal, Dr. Subhash Bhargav Lane,
Opposite Delhi Medical Association, Daryaganj, New Delhi-110002 (INDIA)
Phone: +91-11-43551243; *Mobile:* +91-9871775858, 7678425209
e-mail: katariabook@gmail.com, katariabooks@yahoo.com
Website: www.skkatariaandsons.com

First Edition: August, 2009; Second Edition: July, 2011 Reprint: 2024

Price: ₹ 395/-

Laser Typesetting by : Plus Computers, Delhi

Printed at : New A.S. Printing Press, Delhi

PREFACE

This book **JAVA Programming and Website Design** written to serve as a text book for B.Tech IIIrd Semester students of Guru Gobind Singh Indraprastha University (GGSIPU). There are no prerequisites beyond a general familiarity with the ideas of computers and programs. There is enough material for a full semester of college-level programming.

This covers all features of object oriented programming such as object, classes, inheritance and polymorphism. In this edition we try to cover the feature of Java5. Therefore we have covered new feature of Java5 such as Boxing, enchanced for-loop, object class, instanceof operator and etc.

This book also cover the advance features of Java such as Java applet, Java Beans, Java Swings and JDBC. In each chapter, sufficient worked examples have been given. At the end of each chapter, solved problems and programs have been given for the practice.

Wherever necessary, concepts are given in boxes to facilitate easy grasping and better understanding.

The author will feel highly obliged to all readers for their constructive suggestions healthy criticism of the book which will go a long way in the improvement of the text.

Lalit Arora

ACKNOWLEDGEMENT

It is a pleasure to be able to thank all the people who helped us turn this first edition of Java Programming and Website Design from a list of good ideas into a real book.

We are pleased to express our thanks to the following Academician for allowing us to work under their valuable guidance:

- Dr. Vinod Kumar Sharma, Professor, Gurukul Kangri Vishwavidyalaya, Haridwar
- Dr. Karamjeet Bhatia, Professor, Gurukul Kangri Vishwavidyalaya, Haridwar
- Dr. Rajkumar Bhatia, Professor, Gurukul Kangri Vishwavidyalaya, Haridwar
- Dr. Atul Kumar Jain, Chairman, Ideal Institute of Technology, Ghaziabad.

My special thanks goes to Mr. Ankit Gupta, Sr. Executive, RelSol Pvt. Ltd., Delhi, Mr. Harsh Vardhan, Lecturer, JSS Noida, and Ms. Anjali Gambhir who played a major role in scripting this book.

We are greatly intended to Mr. Sanjeev Kataria and other publishing professionals at **S.K. Kataria and Sons** for their active involvement throughout the production process and for bringing out the book in its present from in record time.

— Lalit Arora

Dedicated to GOD

for blessing me with a son

"MIHIR"

—Lalit Kishore Arora

CONTENTS

Chapter 1

INTRODUCTION

I.I. INTRODUCTION

Java is a powerful object-oriented programming language introduced by Sun Microsystems in 1991. Java was originally called Oak, developed by James Gosling.

Java has built-in support to create programs with a graphical user interface (GUI), utilize the Internet, create client-server solutions, and much more. Programs written in Java can run, without change, on any of the common computer operating systems Windows 95/NT, Macintosh, and Unix. A variant of Java programs called applets can be embedded inside a web page and execute on the computer that is viewing the page, automatically and in a secure environment.

As a language, Java is closely related to C++, which is also object-oriented but retains a lot of idiosyncrasies inherited from its predecessor language C. Java has removed the inconsistent elements from C++, is exclusively object-oriented, and can be considered a modern version of C++.[1] Because of its logical structure Java has quickly become a popular choice as a teaching language, and because of its extensive Internet support and the promise of writing programs once and using them on every operating system Java is becoming more and more accepted in industry.

CHAPTER OUTLINE

[1] Java does have disadvantages. For example, programs written in Java are generally slower than those in C++ and it is difficult to accomplish system-level tasks in Java.

Sun's marketing people refer to all Java versions 1.2 + as JAVA 2 and 1.5 + as JAVA 5.

1.2. FEATURES OF JAVA

The basic features of java are given below by which, java have more fame in the computer world.

1. **Simple:** All the Syntax of java is like C++. There is no preprocessor, and much larger library. Confusing features such as: operator overloading, independent functions, global variables, go-to statements, structure, pointers, are *eliminated* from java.
2. **Object-Oriented:** All the functions are defined inside the classes. You cannot write any independent function. This forces you to think Object Oriented, java also refers as a pure object-oriented language.
3. **Network Savvy:** Java designed with internet in mind. So It supports maximu, network protocol and there component such as switch, socket, port etc. Java standard library provides URL classes.
4. **Platform Independent:** It is platform independent programming language. Java compiles to byte-code (more details section 9.4). This is interpreted (platform = system architecture + operating system).
5. **Robust:** Java is graceful in the presence of software or hardware errors. Java has very good *exception handlers*. Programmers are forced to think about potential sources of error.
6. **Secure:** Java has no pointers. All pointer-related security problems are gone. Java applets have limited access to local machine. It is more secure.
7. **Architecture Neutral:** Java programs can runs on PC, Macintosh, Unix machines, etc. it means system with any architecture of mother board.
8. **Portable:** Java hides architecture-specific concepts. Byte-codes are the same regardless of the machine on which it is run, means you can port your program from one system to other system of different architecture.
9. **High Performance:** Byte-codes are interpreted – making it slow. But we have "Just-in-time" (JIT) compilers. These compile the Byte-codes into the machine code of target machine. This improves the performance.
10. **Multithreaded:** Java programs are capable for easy to set up multiple threads of execution and coordinate parallel processes.
11. **Dynamic:** Because Java programs move through the Internet they can exhibit dynamic behavior.
12. **Internet:** Applets are the most popular programs in java which basically used for Internet web browsers.
13. **Garbage Collection:** Java has an automatic garbage collector which release the objects which are not in use from long time.

1.3. IDENTIFIERS AND VARIABLES

Identifiers refer to the names of variables, functions arrays, classes, etc. created by programmer. They are fundamental requirement of any language. Each language has its own rules for naming these identifiers.

Variable Names

Variable names can start with a letter, '$' (dollar symbol), or '_' (underscore); it cannot start with a number; cannot be a reserved word (or key word).

Method Names

Method name should be verbs or verb phrases with first letter in lowercase, and the first letter of subsequent words capitalized; cannot be reserved words.

Example: setColor()

Class And Interface Names

The class and interface names should be descriptive names that begin with a capital letter, by convention; cannot be a reserved word.

Constant Names

They are in capitals letter.

Example: Font.BOLD, Font.ITALIC

A *variable* is just a memory location (or several locations treated as a unit) that has been given a name so that it can be easily referred to and used in a program.

Variable Declaration

The syntax of variable declaration is as follows :

<datatype> <variable name>

Example: `int num1;`

Like C++ , java allows the declaration of variable anywhere in the scope. This means that a variable can be declare right at the place of its first use.

Variable Initialization

You can initialize the variable at the time of declaration, the syntax of initialization is as follows:

<datatype> <variable name> = value

Example: `double num2 = 3.1419;`

Java Reserved Words

These are explicitly reserved identifiers and cannot be used as names for the program variables or other user-defined program elements. The following table gives the complete set of java keywords.

abstract	default	if	package	this
boolean	do	implements	private	throw
breakdouble	import	protected	throws	
byteelse	instanceof	public	transient	
caseextends	int	return	null	
try Const	for	new	switch	
continue	while		synchronized	super
catchfinal	interface	short	void	
charfinally	long	static	volatile	
classfloat	native			

1.4. DATA TYPES

There are eight so-called data types built into Java. The data types are named byte, short, int, long, float, double, char, and boolean. The first four types hold integers (whole numbers such as 17, -38477, and 0). The four integer types are distinguished by the ranges of integers they can hold. The float and double types hold real numbers (such as 3.6 and -145.99). Again, the two real types are distinguished by their range and accuracy. A variable of type char holds a single character from the Unicode character set. And a variable of type boolean holds one of the two logical values **true** or **false**.

The following table shows the size, default value, minimum value and maximum value of all data types of java.

Data Type	Size	Default	Min Value Max Value
byte (Signe integer)	8 bits	0	–128 +127
short (Signed integer)	16 bits	0	–32,768 +32,767
int (Signed integer)	32 bits	0	–2,147,483,648 +2,147,483,647
long (Signed Integer)	64 bits	0	–9,223,372,036,0 854, 775, 808 +9,223,372,036,0 854, 775, 807
float (IEEE 754 floating-point)	32 bits	0.0	1.4E–45 3.4028235E38
double (IEEE 754 floating-point)	64 bits	0.0	4.9E–324 1.7976931348623157E308
char (Unicode character)	16 bits	\u0000	\u0000 \uFFFF
boolean	1 bit	false	

1.4.1. Reference Type

All data types discussed above, boolean, char, byte, short, int, long, float and double, are called primitive data types. All other types refer to objects rather than primitives. Variables that refer to objects are reference *variables.* For example, you can define a class Person:

```
public class Person
{
}
public class Test
{
    public static void main(String[] args)
    {
        Person emp = new Person();
    }
}
```

The variable emp is a reference variable holding one Person Object.

1.5. OPERATORS

This section gives details of operators in Java. The number of operators in Java is quite large, and we will not cover them all here. Most of the important ones are here; and few will be covered in later chapters as they become relevant.

1.5.1. Arithmetic Operators

Arithmetic operators include addition, subtraction, multiplication, and division. They are indicated by:

Operator	Function
+	Addition
–	Subtraction
*	Multiplication
/	Division
%	Modulus

These operations can be used on values of any numeric type: byte, short, int, long, float, or double. When the computer actually calculates one of these operations, the two values that it combines must be of the same type.

Finally, you might need the *unary minus* operator, which takes the negative of a number. For example, -X has the same value as (-1)*X. For completeness, Java also has a unary plus operator, as in +X, even though it doesn't really do anything.

1.5.2. Increment and Decrement Operators

Adding 1 or Subtracting 1 from or to a variable is an extremely common operation in programming. You might perform the operation of adding 1 to a variable with assignment statements such as:

```
counter = counter + 1;
goalsScored = goalsScored + 1;
```

The effect of the assignment statement x = x + 1 is to take the old value of x, adding 1, and store the answer as the new value of x. The same operation can be accomplished by writing x++ (or, if you prefer, ++x). This actually changes the value of x, so that it has the same effect as writing "x = x + 1". The two statements above could be written

```
counter++;
goalsScored++;
```

Similarly, you could write X-- (or --x) to subtract 1 from x. That is, x-- performs the same computation as x = x - 1.

Adding 1 to a variable is called *incrementing* that variable, and subtracting 1 is called *decrementing*. The operators ++ and -- are called the increment operator and the decrement operator, respectively.

Usually, the operators ++ or --, are used in statements like "x++;" or "x--;". These statements are commands to change the value of x. However, it is also legal to use x++, ++x, x--, or --x as expressions, or as parts of larger expressions. That is, you can write things like:

```
y = x++;        // called post-increment
y = ++x;        // called pre-increment
```

The statement "y = x++;" has the effects of adding 1 to the value of x and, in addition, assigning some value to y. The value assigned to y is the value of the expression x++, which is defined to be the **old** value of x, before the 1 is added. Thus, if the value of x is 6, the statement "y = x++;" will change the value of x to 7, but it will change the value of y to 6 since the value assigned to y is the *old* value of x. On the other hand, the value of ++x is defined to be the **new** value of x, after the 1 is added. So if x is 6, then the statement "y = ++x;" changes the values of both x and y to 7.

The decrement operator, --, works in a similar way for post and pre-decrement.

1.5.3. Relational Operators

Relational operators are used to test whether two values are equal, whether one value is

greater than another, and so forth. The relation operators in Java are:

Operator	Function
==	Equality
!=	Not Equal
<	Less than
>	Greater than
<=	Less than equal to
>=	Greater than equal to

Meaning of these operators are:

A == B Is A "equal to" B?

A != B Is A "not equal to" B?

A < B Is A "less than" B?

A > B Is A "greater than" B?

A <= B Is A "less than or equal to" B?

A >= B Is A "greater than or equal to" B?

These operators can be used to compare values of any of the numeric types. They can also be used to compare values of type char.

1.5.4. Boolean Operators

In Java, the boolean operators shown in following table :

Operator	Function
&&	AND
||	OR
!	NOT

These operators combine two boolean values and the result is also a boolean value. The result tables for these operators are given below :

Table for AND operator (&&)

A && B		Result
True	True	True
True	False	False
False	True	False
False	False	False

Table for OR operator (||) :

A || B		Result
True	True	True
True	False	True
False	True	True
False	False	False

Table for NOT operutor (!) :

! A	Result
True	False
False	True

1.5.5. Conditional Operator

Java has the conditional operator. It's a ternary operator — that is, it has three operands — and it comes in two pieces, ? and :, that have to be used together.

It takes the form

<boolean-expression> ? <expression-1> : <expression-2>

The computer tests the value of *boolean-expression.* If the value is true, it evaluates *expression 1;* otherwise, it evaluates *expression-2.*

For example:

```
next = (N % 2 == 0) ? (N/2) : (3*N+1);
```

will assign the value N/2 to next if N is even (that is, if N % 2 == 0 is true), and it will assign the value (3*N+1) to next if N is odd.

1.5.6. Assignment Operators and Type-Casts

You are already familiar with the assignment statement, which uses the symbol "=" to assign the value of an expression to a variable.

In java following assignment operators are available:

Operator	Meaning
=	Assign right side value to the left side variable
+=	a += b means a = a + b
-=	a -= b means a = a - b
*=	a *= b means a = a * b
/=	a /= b means a = a / b
%=	a %= b means a = a % b

In general, the type of the expression on the right-hand side of an assignment statement must be the same as the type of the variable on the left-hand side. However, in some cases, the computer will automatically convert the value computed by the expression to match the type of the variable. Consider the list of numeric types: byte, short, int, long, float, double. A value of a type that occurs earlier in this list can be converted automatically to a value that occurs later.

For example:

```
int A;
double X;
short B;
A = 17;
X = A;      // OK; value of A is converted to a double
B = A;      // illegal; no automatic conversion
            // from int to short
```

The idea is that conversion should only be done automatically when it can be done without changing the semantics of the value. Any int can be converted to a double with the same numeric value. However, there are int values that lie outside the legal range of shorts. There is simply no way to represent the int 100000 as a short, for example, since the largest value of type short is 32767. In some cases, you might want to force a conversion that wouldn't be done automatically. For this, you can use what is called a *type cast.* A type cast is indicated by putting a type name, in parentheses, in front of the value you want to convert.

For example,

```
int A;
short B;
A = 17;
B = (short)A;          // OK; value of A is explicitly type cast
                       // to a value of type short
```

You can do type casts from any numeric type to any other numeric type. However, you should note that you might change the numeric value of a number by type-casting it.

For example, (short)100000 is -31072.

(The 34464 is obtained by taking the 4-byte int 100000 and throwing away two of those bytes to obtain a short — you've lost the real information that was in those two bytes.)

1.5.7.NPrecedence Rules

If you use several operators in one expression, and if you don't use parentheses to explicitly indicate the order of evaluation, then you have to worry about the precedence rules that determine the order of evaluation. If two same precedence operators occur at same place then order of evaluation will depend on associativity.

Here is a table of the operators, listed in order from highest precedence (evaluated first) to lowest precedence (evaluated last) and associativity:

Operator type	Operators	Associativity
Unary operators	++, (--), !, unary – unary +, type-cast	Right to left
Multiplication and division	*, /, %	Left to right
Addition and subtraction	+, -	Left to right
Relational operators	<, >, <=, >=	Left to right
Equality and inequality	==, !=	Left to right
Boolean and	&&	Left to right
Boolean or	||	Left to right
Conditional operator	?:	Left to right
Assignment operators	=, +=, -=, *=, /=, %=	Right to left

1.6. STATEMENTS IN JAVA

Like C and C++, Java has all three type of statements for creating the programs. The types of statements are :

1. Simple Statements
2. Compound Statements
3. Control Statements

1.6.1. Simple Statements

A simple statement is any expression that terminates with a semicolon. For example

```
var1 = var2 + var3;
var3 = var1++;
var3++;
```

1.6.2. Compound Statements

Related statements can be grouped together in braces to form a compound statement or block. For example:

```
{
    boolean flag = true;
    System.out.println(" i = " + i);
    i++;
}
```

Semantically, a block behave like a statement and can be used anywhere a single statement is allowed. There is no semicolon after the closing braces. Any variable declared in a block remain in scope up to the closing brace. Once the block is exited, the block variables cease to exist. Blocking improves readability of program code and can help make your program easier to control and debug.

1.6.3. Control Statements

Again control statements are divided into three statements.

1. Conditional statements: if, switch
2. Looping statements : for, while, do-while
3. Jumping statements : continue, break

The if statement

The if statement is implemented in two forms:

- Simple **if** statement
- **if.....else** statement

Simple if statement

```
if (expression is true)
{
    action 1;
}
    action 2;
    action 3;
```

if.....else statement

```
if (expression is true)
{
    action 1;
}
else
{
    action 2;
}
    action 3;
```

The switch statement

This is a multiple-branching statement where, based on condition, the control is transferred to one of the many possible points. This is implemented as follows :

```
switch ( expression)
{
    case 1 :
        {
            action 1;
        }
    case 2 :
        {
```

```
            action 2;
        }
    case 3 :
        {
            action 3;
        }
    default :
        {
            action 4;
        }
}
action 5;
```

The do-while statement

The **do-while** is an exit-controlled loop. Based on a condition, the control is transferred back to a particular point in the program. The syntax is as follows :

```
do
{
    action 1;
}
while(condition is true);
```

The while statement

This is also a loop structure, but is an entry-controlled one. The syntax is as follows :

```
while(condition is true)
{
    action 1;
}
action 2;
```

The for statement

The **for** statement is an entry-controlled loop and is used when an action is to be repeated for a predetermined number of times. The syntax is as follows :

```
for(initial value; condition; expression)
{
    action 1;
}
action 2;
```

New form of for loop is introduced in Java 5. The current for loop construct is powerful, but somewhat ill-suited to iteration over collections (collections discussed in chapter10). The standard idiom to iterate over a collection is containing too many words, such as:

```
for (Iterator i = c.iterator(); i.hasNext(); ) // No ForUpdate
    {
        String s = (String) i.next();
```

```
        ...
    }
```

The situation is made little better with the addition of generics:

```
for (Iterator<String> i = c.iterator(); i.hasNext(); ) {
        String s = i.next();
        ...
}
```

The second form of the for loop specifically designed for iteration over collections and arrays. Some time this loop is refer as a foreach loop. Under the syntax of this, the above code could be replaced by this:

```
for (String s : c).
{
        ...
}
```

Similarly, the following code could be used to calculate the sum of an int array.

```
int sum = 0;
for (int j : a)
        sum += j;
```

This loop affects only the compiler; it demands no support from the VM. It does not require a new keyword and is fully compatible with all preexisting programs.

The break statement

The **break** statement can be used to exit any kind of loop when a specific condition is met. Control is transferred immediately to the first statement following the closing brace of the loop body. The use of break statement is useful where you need to abort a loop if some even occurs while executing the loop body. The break statement is also used in switch statements to prevent falling through to a subsequent case label.

The continue statement

The **continue** statement can be used to transferred control to the end of the loop body, prior to the reevaluation of the text expression. If continue is used in used in for loop, the iteration expression is executed before the re-evaluation of the test.

SUMMARY

This chapter gives you the introduction of the java, with its history, and features

- The main feature of java are object-oriented, platform independent, Robust, recure, Portable, and Multithreading.
- Java has all the features of object oriented programming like inheritance, abstraction, encapsulation and polymorphism.
- Java is platform independent, means java programs can be execute under the environment of any operating system.
- Java has no pointers, that why it is called secure.
- Java can set multiple threads in a single program.
- We have also given the method of defining identifiers and variables.
- We have also given a short introduction of operators in this chapter.
- If you are a programmer or student who already known that C & C++, will be filmier with these operators.

— Java also contain that all three types of statements like C and C++ but it removes the 'goto' statements.

SOLVED PROBLEMS

Q.1.1. Consider the following code :

arrayOfInts[j] > arrayOfInts[j+1]

What operators does the code contain?

Ans. [], >, [], +

Q.1.2. Consider the following code :

```
int i = 10;
int n = i++%5;
```

What are the values of i and n after the code is executed?

Ans. i is 11, and n is 0.

Q.1.3. What are the final values of i and n if instead of using the postfix increment operator (i++), you use the prefix version (++i))?

Ans. i is 11, and n is 1.

Q.1.4. To invert the value of a boolean, which operator would you use?

Ans. The logical complement operator "!".

Q.1.5. Which operator is used to compare two values, = or == ?

Ans. The == operator is used for comparison, and = is used for assignment.

Q.1.6. Explain the following code sample: result = someCondition ? value1 : value2;

Ans. This code should be read as: "If someCondition is true, assign the value of value1 to result. Otherwise, assign the value of value2 to result."

Q.1.7. What is the meaning of "final" keyword?

Ans. final keyword is used to declare the variable, method and class. A final variable is a variable whose value cannot be changed once set. A final method cannot be overridden. A final class cannot be subclassed further.

Q.1.8. Why java is called 100% pure object oriented language?

Ans. Java is 100% pure object oriented language because java obeys all the principles of OOAD(Object oriented analysis and design). We can't write java program without using at least one class or object.

Q.1.9. Does JAVA has goto?

Ans. No, Java does not have goto statements Because :

1. goto statements form infinite loops which are drawbacks in a program.
2. goto statements decrease the reliability of a program.
3. goto statements make documentation of a program very difficult. (Documentation means preserving a copy of a program for future use.)

Q.1.10. What are the disadvantages/Limitations of Java?

Ans. Limitations of java:

1. Development time is much longer than using scripting languages.
2. It's a bit slower than C.
3. String manipulation in Perl takes much less lines of code.

EXERCISES

Q.1.1 Who is the developer of Java? And what is the original name of Java?

Q.1.2. What are the main features of Java programming language?

Q.1.3. What do you man by identifiers? And what are the rules for identifiers?

Q.1.4. What the data types are available is Java programming language?

Q.1.5. What are the operators available in Java programming language?

Q.1.6. What do you mean by precedence rules of operator? Give an example.

Q.1.7. Explain Type-Casting . And how it is done in Java programming language?

Q.1.8. What are the statements are available is Java programming language?

Q.1.9. What is the role of conditional operator? Give an example.

Q.1.10. What do you mean by pure object oriented language? Is Java pure object oriented language? If yes the How?

Q.1.11. Is Java is similar to C++ ? give the advantages of Java over C++.

Q.1.12. What is the difference between simple and compound statements?

Q.1.13. Give the syntax of following statements:

(*i*) while (*ii*) do-while

(*iii*) for (*iv*) switch

(*v*) if-else

Q.1.14. Differentiate the following statements:

(*a*) while and do-while (*b*) for and while

(*c*) continue and return

Q.1.15. What is literal?

Q.1.16. Give the meaning of each of the following Java operators:

(*a*) ++ (*b*) &&

(*c*) !=

Q.1.17. What does the computer do when it executes a variable declaration statements. Give an example.

Q.1.18. Explain what is meant by an assignment statement, and give an example. What are assignment statement used for?

Q.1.19. One of primitive type in java is boolean. What is the boolean type? Where are boolean values used? What are its possible values?

Q.1.20. Write a for loop that will print out all the multiple of 3 from 3 to 36, that is 3, 6, 9, 12, 15, 18, 21, 24, 27, 30, 33, 36.

Q.1.21. Show the exact output that would be produced by the following main () routine:

```
public static void main (String [] args)
{
      int x, y;
      x = 5;
      y = 1;
      while (x > 0)
      {
            x = x - 1;
            y = y*x;
            System.out.println (y);
      }
}
```

Chapter 2

Beginning with Java Programs

CHAPTER OUTLINE

2.1 GUIDE LINES FOR JAVA PROGRAMS

Java is an Object-Oriented Programming, all functions are defined in the class and every Java program must follow these guidelines.

- Java is case sensitive, i.e. the word *Program* is different from *program.*
- Curly brackets { and } are used to group statements together.
- An executable Java program must contain at least the following lines as a framework:

```
public class Name
{
    public static void main(String
    args[])
        {
            ... program code ...
        }
}
```

- Every statement whose next statement is not a separate group must end in a semicolon.
- A Java program containing the above framework must be saved using the filename *Name.java*, where *Name* (including correct upper and lower cases) is the word that follows the keywords public class and the file extension is *.java.*

2.2. OBJECTS, CLASSES AND METHODS

Java is an object oriented programming, all programs are based on concept of objects and classes.

2.2.1. Object

Objects are the basic run time entities in an object oriented system. They may represent a person, a place, a bank account, a table of data or any item that the program has to handle. They may also represent user-defined data such as vectors, time and lists. Objects take up space in the memory and have an associated address like a record in Pascal, or a structure in C.

When a program is executed, the objects interact by sending messages to one another. For example, if "customer" and "account" are two objects in a program, then the customer object may send a message to account object requesting for the bank balance. Each object contains data, and code to manipulate the data. Objects can interact without having to know details of each other's data or code. It is sufficient to know the type of message accepted , and the type of response returned by the object.

Objects have unique identity; even if two or more objects have identical attribute values, the system always treats them as different objects. When you create objects in software, you are given a handle to the objects. In java this handle is the name of the objects. Irrespective of what the handle is , it always uniquely identifies the object.

2.2.2. Class

We just explained that objects contain data, and code to manipulate that data. The entire set of data and code of an object can be made a user-defined data type with the help of a *class*. Once a class has been defined, we can create any number of objects belonging to that class.

> **Concept**
>
> *A class is a user defined data type, and the object are also known as class variables.*

A class is a template or blueprint used in creating multiple objects. A class encapsulates all of the data and behaviors that make up an object. When you ask a java to create or instantiate an object, java uses the class definition as a template for building the new object. Each object is associated with the data of type class with which they are created. A class is thus a collection of objects of similar type. For example, apple, orange and mango are members of the class fruit. Classes are user defined data types and behave like the built-in types of programming language.

Class Declaration

A class must be declared using the keyword class followed by the class name.

Syntax

```
class <classname>
{
      — Body of the class
}
```

A typical class declaration is as follows:

```
< access modifier> <classname>
{
      — Member variable declarations;
      — Method declarations and definitions
}
```

2.2.3. Method

When you defines a class for an object-oriented program, you implement all of the behavior of that class in one or more methods. A java method is equivalent to a function, procedure, or subroutine in other languages, except that it must be defined inside a class definition. In other words, there is no support for global methods in java; every method must be defined within a class.

> **Concept**
>
> *A java method is equivalent to a function, procedure, or subroutine in other language, but it must be defined within a class.*

Method declarations

```
<access modifier> <static/final> <return type> <method name>
<arguments list>
{
- - Method body;
}
```

Example:

```
public static void main(String args[])
{
}
```

2.2.4. Access Modifiers

Java controls access to variables, methods and classes through the use of access modifiers. The access modifiers are **private, public** and **protected**.

A **public** is the least restrictive access modifier. Variables and methods declared as public may be seen and used by any other class.

If an access modifier is not specified (default), the variables and methods may be used by any other class within the same package (a container of classes).

Variables and methods declared as **protected** may be seen from any subclass of that class. It may also be seen from any class with the package in which it exists.

The most restrictive access modifier is **private**. A private method or variable may not be accessed by any other class. The restriction applies to all other classes regardless of their package.

2.3. A SIMPLE PROGRAM

To create a Java program you first create a text file containing the lines

```
public class Name
{
      public static void main(String args[])
      {
```

```
        ... more lines ...
    }
}
```

The file containing our code is called the source code file. This source code file can be create any text editor like notepad, MS-Word, WordPad or DOS Editor. But java source code files should be saved as *Name.java*, where *Name* is the name that appears in the first line of the program: *public class Name*. That *Name* is referred to as the name of the class, or program. By convention its first letter is capitalized.

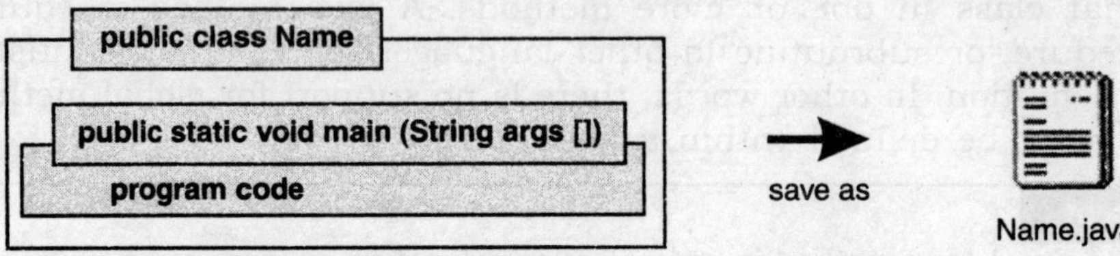

Fig. 2.1. Saving a Java source code file

Example 2.1: The first Program

Create a program containing the necessary Java code to get the computer to write on the screen.

Our first Java program looks as follows:

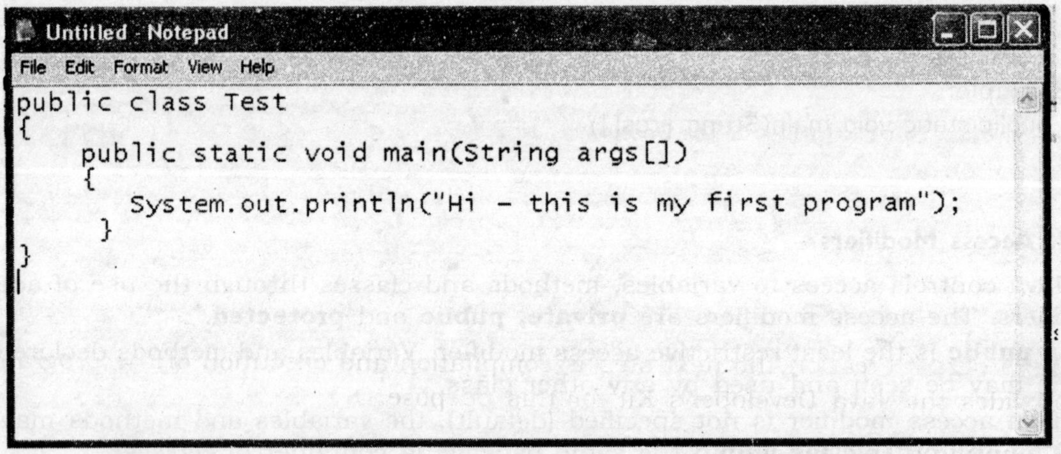


```
public class Test
{
    public static void main(String args[])
    {
        System.out.println("Hi - this is my first program");
    }
}
```

This program, or class, is called and must be saved under the file name Test.java*, you should create a folder for java programs like as :

* You can save file with other name also but at run time you should call the class name for execution.

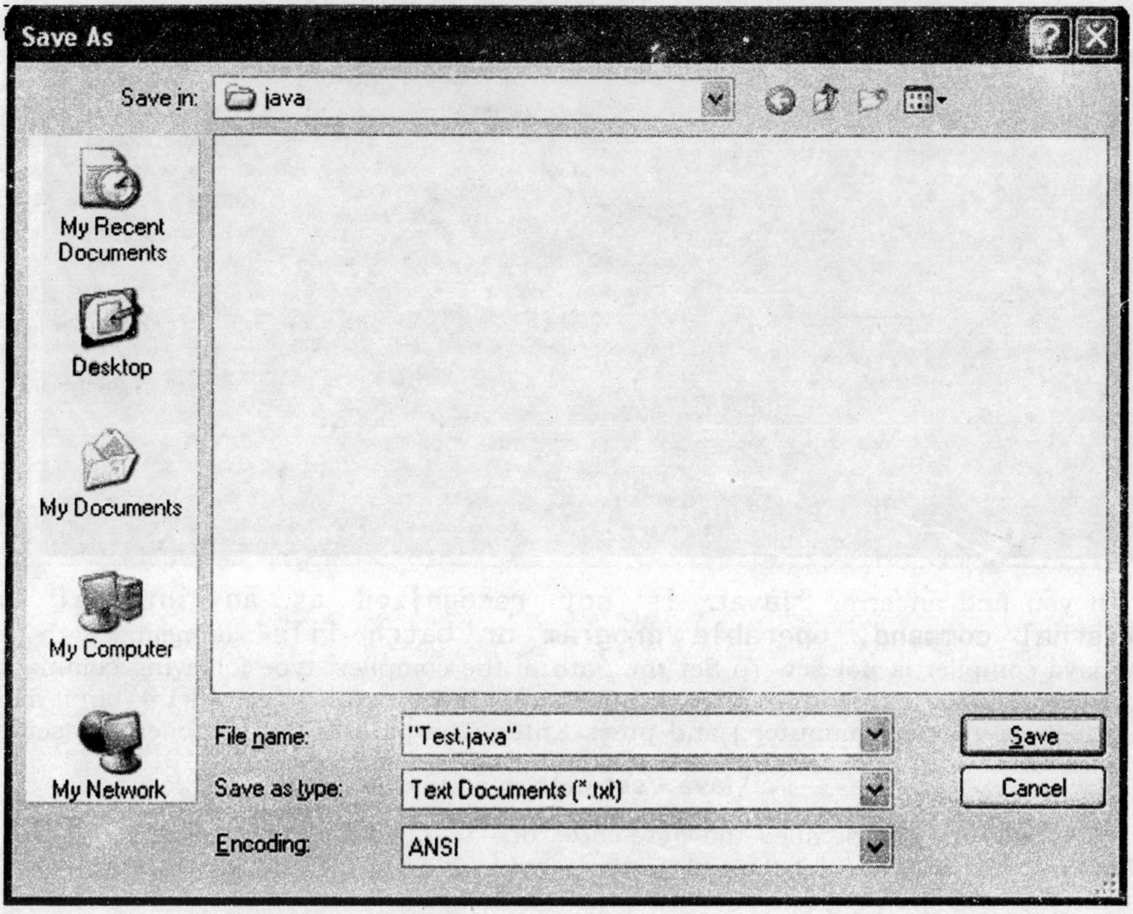

2.4. COMPILING AND EXECUTION

After the program creation the next step is compilation and execution of the program. Java provides the Java Developer's Kit for this purpose.

2.4.1. Compiling a Java Program

Compiling is the process of transforming the source code file into a format that the computer can understand and process. To compile the program,

- Select **Start → Programs → MS-DOS Prompt**
- Type the command " **cd c:\java"**
- Type **"javac Test.java"** and Press Enter

For example,

If you find an error **"javac is not recognized as an internal or external command, operable program or batch file"**. It means path for the java compiler is not set. To Set the path of the compiler, type following command (or wherever javac.exe is located, according to my computer it is in c:\ jdk1.4\bin it may be different in your computer) and press Enter. The path of the compiler will set.

```
c:\java> set path = c:\jdk1.4\bin
```

The resulting file is called the ***byte-code***, or ***class file***. The name of the class file is the same as the name of the program plus the extension *.class*. The program *javac* from the Java Developer's Kit is used to transform a source code file into a class file.

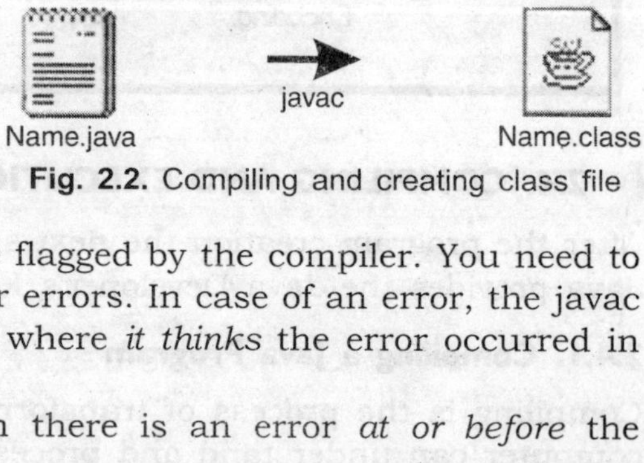

Fig. 2.2. Compiling and creating class file

If a source code contains any errors, they are flagged by the compiler. You need to fix them and re-compile until there are no further errors. In case of an error, the javac compiler shows the line number and position of where *it thinks* the error occurred in your source code.

- If the compiler points out an error, then there is an error *at or before* the indicated position.
- If the compiler reports a certain number of errors, than this is *the least* amount of errors.
- If one error is fixed, other errors may automatically disappear or new ones may appear.

Fix your source code a few errors at a time. Recompile often to see if the number of errors and the error messages change until no errors are reported. If you can not find an error at the position indicated by the compiler, look at the code *before* that position

2.4.2. Executing a Java Program or Class

The Java compiler does not produce an executable file, so Java programs can not execute under the operating system of your machine. Instead they execute inside a *Java Virtual Machine*, which is invoked using the java program of the JDK.

To execute a Java program the Java Developer's Kit provides a program called 'java'. When executing that program with your class file as parameter the following happens:

- the Java Virtual Machine (JVM) must be present on your computer
- the JVM locates and reads your class files
- the JVM inspects your class file for any security violations
- the JVM executes, or interprets, your class file according to its instructions if possible

Under Windows and Unix, execute a program by typing at the command prompt, where is the name of the program (no extension). For example,

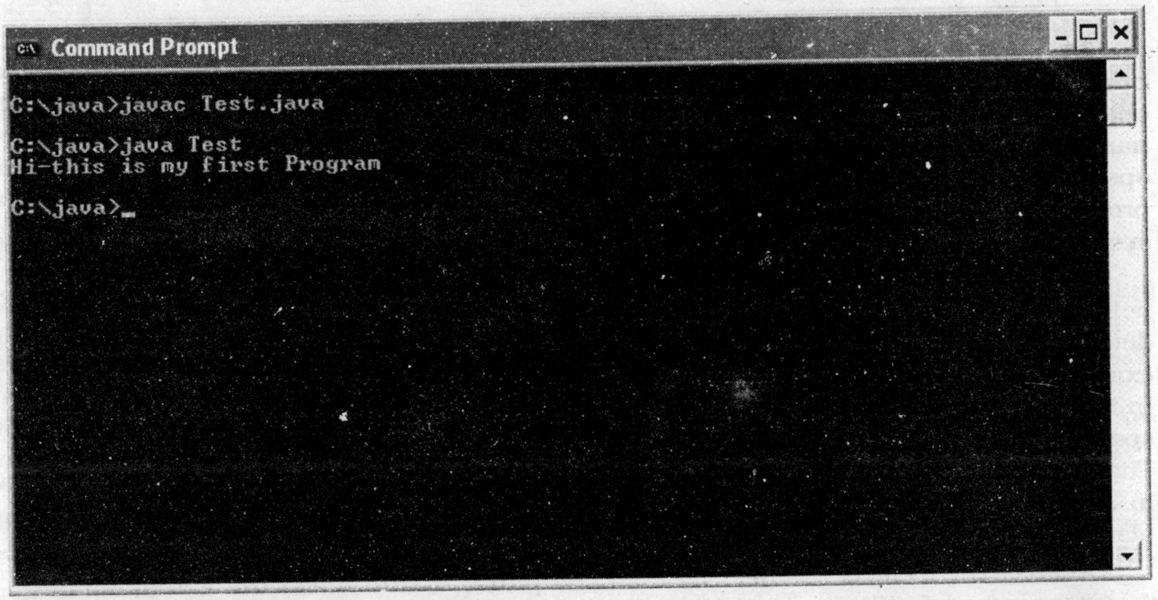

On a Macintosh, double-click the icon and select the appropriate class file.

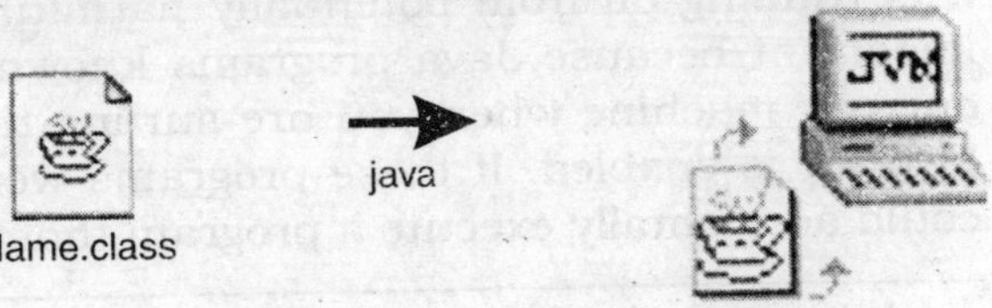

Fig. 2.3. Executing a class file

A good question at this point is which line in a Java program executes *first*. The default program entry point is that part of a class (or program) where execution begins. For every Java class (or program), the standard program entry point consists of the line:

```
public static void main (String args [])
```

If that line is not present in your source code, the JVM can not execute your program and displays an error message.

2.5. JAVA VIRTUAL MACHINE (JVM)

The Java Virtual Machine (JVM) is a platform-independent engine used to run Java applets and applications. The JVM knows nothing of the Java programming language, but it does understand the particular file format of the platform and implementation independent class file produced by a Java compiler. Therefore, class files produced by a Java compiler on one system can execute without change on any system that can invoke a Java Virtual Machine.[1]

When invoked with a particular class file, the JVM loads the file, goes through a verification process to ensure system security, and executes the instructions in that class file.

> **Concept**
>
> *The Java Virtual Machine (JVM) is a platform-independent engine used to run Java applets and applications.*

The JVM, in other words, forms a layer between the operating system and the Java program that is trying to execute. That explains how *one* Java program can run without change on a variety of systems: it can not! A Java program runs on only *one* system, namely the **Java Virtual Machine**. That virtual system, in turn, runs on a variety of operating systems and is programmed quite differently for various systems. To the Java programmer, it provides a unified interface to the actual system calls of the operating system.[2]

You can include graphics, graphical user interface elements, multimedia, and networking operations in a Java program and the JVM will negotiate the necessary details between the class file(s) and the underlying operating system. The JVM produces exactly the same results – in theory – regardless of the underlying operating system. In the Basic (or C, or C++) programming language, for example, you can create code that specifies to multiply two integers 1000 and 2000 and store the result as another integer. That code works fine on some systems, but can produce negative numbers on others.[3] In Java, this can not happen: either the code fails on all platforms, or it works on all platforms.

Because the JVM is in effect its own computer, it can shield the actual computer it is running on from potentially harmful effects of a Java program. This is especially important because Java programs known as *applets* can *automatically* start executing on your machine when you are surfing the web if the appropriate feature of your web browser is enabled. If these programs were allowed to interfere with your system, you could accidentally execute a program that would proceed to erase your entire disk. That,

---

1 In general, the Java Virtual Machine is an abstractly specified class file interpreter that can be realized by different software makers. The JVM that comes with the JDK was created by SUN Microsystems, but any other JVM is also able to run the same class files. Different JVM's can vary in efficiency, but they all must run the same class files.

2 This is somewhat similar to old Basic programs: a simple Basic program can run on virtually any system that has a Basic interpreter installed since the interpreter mediates between the program trying to run and the operating system.

3 Programming languages have a *largest possible* integer whose value can differ on different systems. A C++ program executing on a machine with a largest integer bigger than 2,000,000 produces the correct result, but on a system where the largest integer is, say, 32,767 it fails. The JVM has the *same* largest integer on every platform.

of course, would prompt people to disable Java on their web browser, which in turn would be bad news for anyone who supports the Java concept.

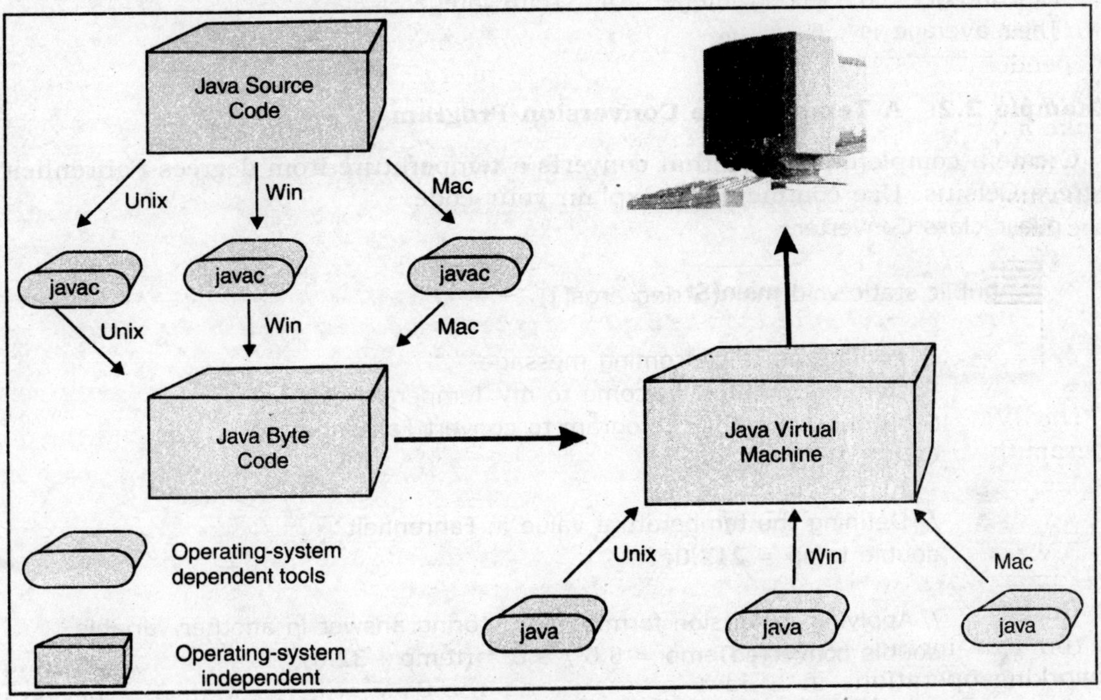

Fig. 2.4. Illustrating the machine dependent/independent parts of Java programs

2.6. SOME EXAMPLE OF PROGRAMS

Example 2.1: A Simple Average Program

Create a simple program to compute the average of 3 positive integers.

```
public class ComputeAverage
{
    public static void main(String[] args)
    {
        int a, b, c ;
        int sum, avg ;
        a = 6;
        b = 4;
        c = 8;
        sum = a + b + c ;
        avg = sum / 3;
        System.out.println("The values of the positive integers.");
        System.out.println("First integer : " + a + ", Second integer
        : " + b + ", Third integer : " + c);
        System.out.println ("Their average is :" + average + ".");
    }
}
```

OUTPUT:

```
The values of the positive integers.
First integer : 6, Second integer : 4 , Third integer : 8
Their average is : 6
```

Example 2.2: A Temperature Conversion Program

Create a complete program that converts a temperature from degrees Fahrenheit in degrees Celsius. Use comments to explain your code.

```
public class Converter
{
    public static void main(String args[])
    {
        // Printing out a welcoming message
        System.out.println("Welcome to my Temperature Converter.");
        System.out.println("\nProgram to convert Fahrenheit to
        Celcius.\n");

        // Defining the temperature value in Fahrenheit
        double temp = 212.0;

        // Applying conversion formula and storing answer in another variable
        double convertedTemp = 5.0 / 9.0 * (temp - 32.0);

        // Printing out the complete answer
        System.out.println(temp + " Fahrenheit = " + convertedTemp
        + " Celcius.");
    }
}
```

OUTPUT:

```
Welcome to my Temperature convertor.
Program to convert Fahrenheit to celcius.
212.0 Fahrenheit = 100 Celcius
```

Example 2.3: A Simple interest Program

Create a simple program to compute the amount of interest that is earned on Rs. 17,000 invested at an interest rate of 0.07 for one year.

```
public class Interest
{
    public static void main(String args[])
    {

        double principal; // The value of the investment.
        double rate;          // The annual interest rate.
        double interest; // Interest earned in one year.
```

```
principal = 17000;
rate = 0.07;
interest = principal * rate;    // Compute the interest.
principal = principal + interest;

// Compute value of investment after one year, with interest.
// (Note: The new value replaces the old value of principal.)

System.out.println("The interest earned is Rs. " + interest);
System.out.println("The value of the investment after one
year is Rs. " + principal);

}
}
```

OUTPUT:

The interest earned is Rs. 1190

The value of the investment after one year is Rs. 18190s

Example 2.4. A Simple interest Program for 5 years

Create a simple program to compute the amount of interest that is earned on an investment over a period of 5 years. The initial amount of the investment is Rs. 17,000 and the interest rate is 0.07. Display the value of the investment at the end of each year.

```
public class Interest3
{
    public static void main(String args[])
    {
        double principal=17000; // The value of the investment.
        double rate=0.07;               // The annual interest rate.

        /* Simulate the investment for 5 years. */
        int years; //Counts the number of years that have passed.
        years = 0;
        while (years < 5)
        {
        double interest; // Interest for this year.
        interest = principal * rate;
        principal = principal + interest; // Add it to principal.
        years = years + 1;              // Count the current year.
        System.out.println("The value of the investment after "
        + years);
        System.out.print(" years is Rs.+ principal);
        }
    }
}
```

OUTPUT:

```
The value of the investment after 1 year is Rs. 18190
The value of the investment after 2 year is Rs. 19463.3
The value of the investment after 3 year is Rs. 20825.73
The value of the investment after 4 year is Rs. 22283.53
The value of the investment after 5 year is Rs. 23843.4
```

SUMMARY

This chapter is very important chapter for every Beginner of Java language. In this we have explained the way of programming. We have also given the pictures of required steps which will helpful for every programmer.

— You must save your program with class name, and extension must be "Java".

— Java has a new concept of Byte code, which is intermediate level code and interpreted by JVM.

— JVM (Java Virtual Machine) is a software used to run java applications and applets.

— We have also explained the whole procedure of compilation and execution of the java programs.

SOLVED PROBLEMS

Q.2.1. What is BYTE Code?

Ans. A BYTE Code is intermediate level code, which is interpreted by the JVM. It is not directly executable on the machine. A BYTE code is a mediator code between java program and machine code of Operating System (O.S) and is available in *'.class'* file. A *'.class'* file is generated when the java program or application gets compile. Byte code is same to any O.S but machine code varies to every O.S.

Q.2.2. What is the Java Virtual Machine?

Ans. The Java Virtual Machine is a software that can be ported onto various hardware-based platforms.

Q.2.3. What gives java it's "write once and run anywhere" nature?

Ans. When java program is written and compiled then it will create a (*.class*) file which consists of byte code instructions, underastandable to JVM. This class file is system independent. Every system has it's own JVM. so JVM will convert this byte code into machine language understandable to that system. So it has run write once and run anywhere nature. Java is 'platform indipendent' programing language. The Byte code is same for any Operating System. That's why we can execute an application in one O.S which was already compiled in other O.S, we can directly copy the application from one OS to Other OS with out writing again.

Q.2.4. What is native code?

Ans. The native code is code that after you compile it, the compiled code runs on a specific hardware platform.

Q.2.5. What is a Java compiler?

Ans. A Java compiler is a program which converts **Java source code** into Java bytecode.

A basic Java compiler is included as part of the JDK (Java Development Kit). This Java compiler is called "javac".

Q.2.6. What is the JRE?

Ans. The JRE is the Java Runtime Environment.

The JRE consists of the Java Virtual Machine, the Java libraries, and all other components necessary to run Java applications and applets.

Sun distributes the JRE seperately, or as part of J2EE, J2SE, and J2ME.

Q.2.7. What is special about static variables?

Ans. A static variable is shared by all instances of a class. Every instance has a reference to the same variable and can modify it directly. Static instances are declared with the static modifier.

Q.2.8. Write a program that produces the following output:

Hello World!
It's been nice knowing you.
Goodbye world!

Ans. This is three lines of text so it needs three System.out.println() statements. In other words:

```
class HelloWorld {
public static void main (String args[]) {

        System.out.println("Hello World!");
        System.out.println("It's been nice knowing you.");
        System.out.println("Goodbye world!");

        }

}
```

C programmer's can note that \n works in Java just like it works in C, C++ so you could also write

```
class HelloWorld {
        public static void main (String args[]) {

        System.out.println("Hello World!\nIt's been nice knowing
        you.\nGoodbye world!");

        }

}
```

Q.2.9. Write a program that draws the following figures one above the other.

```
* * * * *          *
* * * * *          * *
* * * * *          * * *
* * * * *          * * * *
```

Ans. Now modify it to draw them next to each other like above.

```
class DrawPicture1 {

        public static void main (String args[]) {

                System.out.println("* * * * *");
                System.out.println("* * * * *");
```

```
            System.out.println("* * * * *");
            System.out.println("* * * * *");
            System.out.println(" *");
            System.out.println(" * *");
            System.out.println(" * * *");
            System.out.println(" * * * *");
            }

    }
```

Q.2.10. The Fibonacci numbers are defined as follows. The zeroth Fibonacci number is 1. The first Fibonacci number is also 1. The second Fibonacci number is 1 + 1 = 2. The third Fibonacci number is 1 + 2 = 3. In other words, except for the first two numbers each Fibonacci number is the sum of the two previous numbers. Write a program that prints the first 20 Fibonacci numbers.

Ans.

```
class Fibonacci {

    public static void main (String args[]) {

        int i;
        int fibold;
        int fibnew;
        int temp;

        i = 0;
        fibold = 1;
        System.out.println(fibold);
        fibnew = 1;
        System.out.println(fibnew);

        while (i < 19) {

        temp = fibold + fibnew;
        fibold = fibnew;
    fibnew = temp;
    System.out.println(fibnew);
    i = i + 1;
    }

  }

}
```

Q.2.11. There are exactly 2.54 centimeters to an inch. Write a program that takes a number of inches from the command line and converts it to centimeters.

Ans.

```
class intocm {

    public static void main (String[] args) {
```

```
        double inches = Double.valueOf(args[0]).doubleValue();
        double cm = inches * 2.54;
        System.out.println(cm + " centimeters");

    }

}
```

Q.2.12. Write the inverse program that reads a number of centimeters from the command line and converts it to inches.

Ans.

```
class cmtoin {

    public static void main (String[] args) {

    double cm = Double.valueOf(args[0]).doubleValue();
    double inches = cm / 2.54;
    System.out.println(inches + " inches");

    }

}
```

Q.2.13. There are 454 grams in a pound and 1000 grams in a kilogram. Write programs that convert pounds to kilograms and kilograms to pounds. Read the number to be converted from the command line. Can you make this one program instead of two?

Ans.

```
class lbstokg {

    public static void main (String[] args) {

        double lbs = Double.valueOf(args[0]).doubleValue();
        double kg = lbs * 454.0/1000.0;
        System.out.println(kg + " kg");

    }

}
class kgtolbs {

    public static void main (String[] args) {

        double kg = Double.valueOf(args[0]).doubleValue();
        double lbs = kg * 1000.0/454.0;
        System.out.println(kg + " kg");

    }

}
```

to make this one program you need a way of telling whether the number entered on the command line is in pounds or kilograms. Let's adopt the convention that if the number is given in kilograms, then the second command line argument will bekg. Thus

```
class lbkg {

    public static void main (String[] args) {

    if (args[1].equals("kg")) {
    double kg = Double.valueOf(args[0]).doubleValue();
    double lbs = kg * 1000.0/454.0;
    System.out.println(kg + " kg");
  }
  else {
    double lbs = Double.valueOf(args[0]).doubleValue();
    double kg = lbs * 454.0/1000.0;
    System.out.println(kg + " kg");
  }

  }

}
```

Q.2.14. The equivalent resistance of resistors connected in series is calculated by adding the resistances of the individual resistors. Write a program that accepts a series of resistances from the command line and outputs the equivalent resistance.

Ans.

```
class seriesResistance {

    public static void main (String[] args) {

        double total = 0;
        double resistance;

        int i = 0;
        while (i < args.length) {
            resistance = Double.valueOf(args[i]).doubleValue();
            total = total + resistance;
            i = i + 1;
        }
        System.out.println("The total resistance is " + total);
    }

}
```

Q.2.15. The formula for resistors connected in parallel is a little more complex. Given two resistors with resistances R1 and R2 connected in parallel the equivalent resistance is given by the inverse of the sum of the inverses, i.e.

If there are more than two resistors you continue to invert the sum of their inverses; e.g. for four resistors you have:

Write a program that calculates the resistance of a a group of resistors arranged in parallel.

Ans.

```
class parallelResistance {

    public static void main (String[] args) {

        double total = 0;
        double resistance;

        int i = 0;
        while (i < args.length) {
            resistance = 1.0/Double.valueOf(args[i]).doubleValue();
            total = total + resistance;
            i = i + 1;
        }
            total = 1.0/total;
            System.out.println("The total resistance is " + total);

    }

}
```

EXERCISES

Q.2.1. What are the main guide lines for java programs?

Q.2.2. How can you define object in your words?

Q.2.3. What is the class? And explain the access modifiers.

Q.2.4. What is the extension of java programs? And what should the name of java program?

Q.2.5. What is the role of JVM? Is it a hardware device?

Q.2.6. What do you mean by byte code?

Q.2.7. What is the difference between private and protected class?

Q.2.8. In Java, classes have two fundamentally different purposes. What are they?

Q.2.9. In you have the source code for a Java program, and you want to run that program, you will need both a compiler and an interpreter. What does the Java compilerdo, and what does the Java interpreterdo?

Q.2.10. Java is an object oriented programming language. What is an object?

Q.2.11. Java is a "platform-independent language". What does this mean?

Q.2.12. Object-oriented programming uses classes and objects. What are classes and what are objects? What is the relationship between classes and objects?

Chapter 3

BASIC JAVA PROGRAMS

CHAPTER OUTLINE

- Simple Programs
- Command Line Arguments
    - Using Characters
    - Using Integers
    - Using Strings
- Console Inputs
    - Scanner Class
    - Buffered Reader Class
    - Some More Input Forms
- Multiple "mail()" Functions
- Multi-file Program

3.1. SIMPLE PROGRAMS

In this section we demonstrate some example programs which are most important for your skills and practice. Here we tried to remove all possible errors, but if you find any error please convey us.

Example 3.1: Program to display the addition, subtraction, multiplication and division of two numbers.

```
class Example
{
public static void main(String[] args)
{
    int a,b;
a=10;
b=4;
System.out.println("Addition ="+ (a+b));
System.out.println("Subtraction ="+ (a-b));
System.out.println("Multiplication ="+ (a*b));
System.out.println("Division ="+ (float)(a/b));
}
}
```

OUTPUT:

```
Addition =14
Subtraction =6
Multiplication =40
Division =2.0
```

Example 3.2: Program to interchange the values of two variables using third variable.

```
class Example
{
public static void main(String[] args)
{
int a,b,c;
a = 10;
b = 4;
System.out.println("Before Interchange , A ="+a+" and B = "+b);
c = a;
a = b;
b = c;
System.out.println("After Interchange , A ="+a+" and B = "+b);
}
}
```

OUTPUT:

```
Before Interchange,     A = 10 and B = 4
After Interchange,      A = 4 and B = 10
```

Example 3.3: Program to interchange the values of two variables without using third variable.

```
class Example
{
public static void main(String[] args)
{
int a, b;
a = 10;
b = 4;
System out.println("Before Interchange , A ="+a+" and B = "+b);
a = a + b;
b = a - b;
a = a - b;
System.out.println("After Interchange , A ="+a+" and B = "+b);
}
}
```

OUTPUT:

```
Before Interchange,     A = 10 and B = 4
After Interchange,      A = 4 and B = 10
```

Example 3.4: Program to find the maximum and minimum number in three given numbers.

```
class example
{
public static void main(String[] args)
{

int a, b, c, max, min;
a = 10;
b = 4;
c = 11;

max = min = a;

if(max<b) max = b;
if(max<c) max = c;

if(min>b) min = b;
if(min>c) min = c;

System.out.println("Maximum Number ="+max);

System.out.println("Minimum Number ="+min);
}
}
```

OUTPUT:

```
Maximum Number = 11
Minimum Number = 4
```

Example 3.5: Program to find the maximum and minimum number in three given numbers without using if statement (Hint: use ternary operator "?:").

```
class example
{
public static void main(String[] args)
{
    int a, b, c, max, min;
    a = 10;
    b = 4;
    c = 11;

    max = (a>b?(a>c?a:c):(b>c?b:c));
    min = (a<b?(a<c?a:c):(b<c?b:c));

    System.out.println("Maximum Number ="+max);
    System.out.println("Minimum Number ="+min);
}
}
```

OUTPUT

```
Maximum number is 11
Minimum number is 4
```

Example 3.6: Program to d isplay the first 10 natural numbers and their sum.

```
class example
{
public static void main(String[] args)
{

int sum=0;
for(int i=1; i<=10; i++)
{
    System.out.println(i);
    sum = sum + i;
}
System.out.println("Sum of above numbers = "+sum);
}
}
```

OUTPUT

```
1
2
3
4
5
6
7
8
9
10
Sum of above numbers = 55
```

Example 3.7: Program to display the sum of first ten even natural number's square (e.g., $2^2 + 4^2 + 6^2 + 8^2 + ...$)

```
class example
{
public static void main(String[] args)
{

int sum=0;
int even=2;
for(int i=1; i<=10; i++)
{
sum = sum +even * even;
even += 2;  //even = even + 2;
```

```
}
System.out.println("Sum of first 10 even natural number's Square = "+sum);
}
}
```

OUTPUT

```
Sum of first 10 even natural number's Square = 1540
```

Example 3.8: Program to display the table of given number.

```
class example
{
public static void main(String[] args)
{
        int num=5;
        System.out.println("Table of "+num+ " is given below");
        for(int i=1; i<=10; i++)
        {
        System.out.println(num +" * "+i+" = "+num*i);
        }
}
}
```

OUTPUT

```
Table of 5 is given below
5 * 1 = 5
5 * 2 = 10
5 * 3 = 15
5 * 4 = 20
5 * 5 = 25
5 * 6 = 30
5 * 7 = 35
5 * 8 = 40
5 * 9 = 45
5 * 10 = 50
```

Example 3.9: Program to print series of prime number up to given number.

```
class example
{
public static void main(String[] args)
{
        int num=20;
        int p;
        System.out.println("The series of prime Numbers");
        for(int i=2; i<=num; i++)
        { p = 1;
                for (int j = 2; j<=i/2; j++)
                {
```

```
        if (i% j==0)
            p = p*0;
        else
            p=p*1;
    }
    if (p==1)
        System.out.print(i+", ");
    }
}
}
```

OUTPUT

```
The series of prime numbers:
2, 3, 5, 7, 11, 13, 17, 19,
```

3.2. COMMAND LINE ARGUMENTS

Command line arguments are parameters supplied to the Main method at the time of invoking it for execution. To understand the concept see the example given below :

Example 3.10: Program to accept two arguments from the command line and display on the console. (Sample.java)

```
class sample
{
public static void main(String[] args)
{

System.out.println("You have entered following arguments");
System.out.println("First Argument  : "+args[0]);
System.out.println("Second Argument : "+args[1]);
}
}
```

EXECUTION COMMAND (where program name is Sample.java)

C:\> java sample **Welcome Home**

OUTPUT

```
You have entered following arguments
First Argument  : Welcome
Second Argument : Home
```

In previous examples of this book, we have used the main method with some parameters. Notice that in the above example how the main is declared.

```
public static void main(String[] args)
```

main is declared with a parameters args. The parameter args is an array of strings. Any arguments provided in the command line at the time of execution, are passed to the array args as its elements. We can access the array elements by using a subscript like args[0], args[1] and so on.

Like as in the command line "C:\> java sample Welcome Home", args[0] and args[1] contains :

args[0] ® Welcome
args[1] ® Home

where program name is → sample

There fore the values will be display of these parameters in System.out.println() statement.

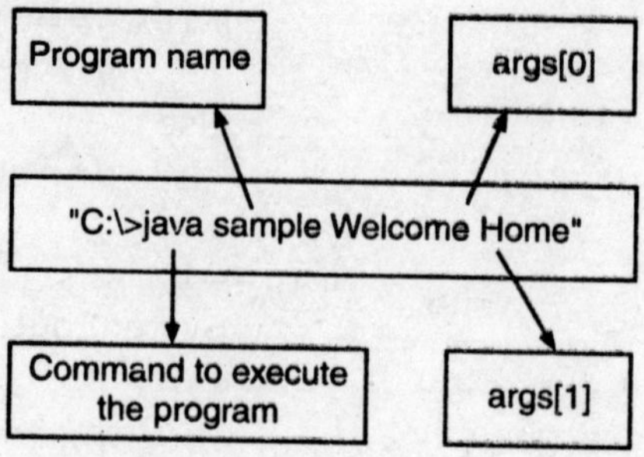

3.2.1. Using Characters

In this section we will handle the command line arguments as a characters. As you saw in previous example that we receive the command line arguments in the form of strings therefore to use arguments as characters you have to convert them into characters. Here we convert the string arguments into character using CharAT() method, which is the method of string class.

Example 3.11: Program to input an alphabet as character and print its predecessor and successor character.

```
class sample
{
public static void main(String[] args)
{
char a=args[0].charAt(0);

System.out.println("You have entered "+a+" Character.");

System.out.println("The Predecessor Character is  : "+(char)(a-1));
System.out.println("The Successor Character is : "+(char)(a+1));
}
}
```

EXECUTION COMMAND (if program name is sample.java):
C:\> java sample D

OUTPUT

```
You Have Entered D Character
The predecessor character is: C
The successor character is: E
```

Example 3.12: Program to input an alphabet character in lower case and display in an upper case or vice-versa.

```
class sample
{
public static void main(String[] args)
{
      char ch1=args[0].charAt(0);
      int ch=(int)ch1;
```

```
        if (ch>=65 && ch<=90)     //ASCII value for A to Z  = 65 to 90
            {
            ch = ch + 32;
            System.out.println("Lower Case = "+(char)(ch));
            }
        else if(ch>=97 && ch<=122) // ASCII value for a to z = 97 to 122
            {
            ch = ch-32;
            System.out.println("Upper Case = "+(char)(ch));
            }
    }
    }
```

Execution Command (if program name is sample.java):

C:\> java sample d

OUTPUT

Upper Case = D

Execution Command (if program name is sample.java):

C:\> java sample N

OUTPUT

Lower Case = n

3.2.2. Using Integers

In this section we will handle the command line arguments as integers. We know that we receive the command line arguments in the form of strings therefore to use arguments as integers you have to convert them into integers. Here we convert the string arguments into integer using Integer.parseInt() method.

Example 3.13: Program to input a number between 0 to 9 and display in words.

```
class example
{
public static void main(String[] args)
{
int a=Integer.parseInt(args[0]);
switch(a)
    {
case 0    :     System.out.println("Zero");
                break;
case 1    :     System.out.println("One");
                break;
case 2    :     System.out.println("Two");
                break;
case 3    :     System.out.println("Three");
                break;
case 4    :     System.out.println("Four");
                break;
case 5    :     System.out.println("Five");
                break;
```

```
        case 6    :  System.out.println("Six");
                     break;
        case 7    :  System.out.println("Seven");
                     break;
        case 8    :  System.out.println("Eight");
                     break;
        case 9    :  System.out.println("Nine");
                     break;
        default   :  System.out.println("Invalid number out of Range");
                     break;
      }
   }
   }
```

Execution Command (if program name is example.java):

C:\> java example 3

OUTPUT

Three

Execution Command (if program name is Example.java):

C:\> java example 12

OUTPUT

Invalid number out of Range

Example 3.14: Program to display the series of numbers given below where n is a row number input by Command line. (input n = 4)

(a)		**(b)**		**(c)**	
	1		**1**		**1**
	1 2		**2 1**		**2 5**
	1 2 3		**3 2 1**		**3 6 8**
	1 2 3 4		**4 3 2 1**		**4 7 9 10**

(a)

```
class example
{
public static void main(String[] args)
{
      int n=Integer.parseInt(args[0]);
      for(int i=1; i<=n; i++)
      {
            for(int j = 1;j<=i; j++)
            {
            System.out.print(" "+j);
            }
      System.out.println();
      }
}
}
```

Execution Command (if program name is example.java):

C:\> java example 4

OUTPUT

```
1
1 2
1 2 3
1 2 3 4
```

(b)

```
class example
{
public static void main(String[] args)
{

      int n=Integer.parseInt(args[0]);
      for(int i=1; i<=n; i++)
      {
            for(int j=i; j>=1; j--)
            {

            System.out.print(" "+j);
            }
      System.out.println();
      }
}
}
```

Execution Command (if program name is example.java):

C:\> java example 5

OUTPUT

```
1
2 1
3 2 1
4 3 2 1
5 4 3 2 1
```

(c)

```
class example
{
public static void main(String[] args)
{
      int n = Integer.parseInt(args[0]);
      int m;
      int x;
      for(int i = 1; i <= n; i++)
      {
            x = 3;
            m = i;
            for(int j = 1; j <= i; j++)
            {
                  if(j==1)
                        System.out.print(" "+m);
```

```
                else
                {
                        m = m + x;
                        System.out.print(" "+m);
                        x--;
                }
            }
            System.out.println();
        }
    }
    }
```

Execution Command (if program name is example.java):

C:\> java example 4

OUTPUT

```
1
2 5
3 6 8
4 7 9 10
```

Example 3.15: Program to input number through command line and check it is prime or not.

```
class example
{
public static void main(String[] args)
{

    int num = Integer.parseInt(args[0]);
    int p = 1;
    for(int i=2;i<=num/2; i++)
    {
        if(num%i==0)
            p = p * 0;
        else
            p = p * 1;
    }
if(p!=0)
        System.out.println("Number is prime");
else
        System.out.println("Number is not prime");
}
}
```

Execution Command (if program name is example.java):

C:\> java example 6

OUTPUT

Number is not prime.

Execution Command (if program name is example.java):

```
C:\> java example 5
```

OUTPUT

```
Number is prime.
```

3.2.3. Using Strings

In this section we will handle the command line arguments as strings. As you know that we receive the command line arguments in the form of strings so there is no need to convert arguments into the string. you can directly use the arguments as strings. Let see some examples.

Example 3.16: Program to input your full name through command line and find the number of words in your name.

```
class example
{
public static void main(String[] args)
{
        int num = args.length;
        System.out.println("Your Name contains "+num+" words");
}
}
```

Execution Command (if program name is example.java):

```
C:\> java example Raj Kumar Bhatia
```

OUTPUT

```
Your Name contains 3 words
```

Example 3.17: Program to input your full name through command line and find the number of characters in your name.

```
class example
{
public static void main(String[] args)
{
        String name=null;
        for(String s : args)
        name = name + s;
        int num = name.length();
        System.out.println("Your Name contains "+num+" characters");
}
}
```

Execution Command (if program name is example.java):

```
C:\> java example Raj Kumar Bhatia
```

OUTPUT

```
Your Name contains 14 characters
```

Example 3.18: Program to input your first name through command line and display in reverse.

```
class example
{
public static void main(String[] args)
{
    String name = args[0];
    char[] arr=new char[10];
    int n=name.length();
    for(int i=0;i<name.length();i++)
    {
        char s=name.charAt(i);
        arr[--n]=s;
    }
    System.out.println("Your Name in Reverse order is: ");
    for(int i=0;i<name.length();i++)
    System.out.print(arr[i]);
}
}
```

Execution Command (if program name is example.java):
C:\> java example Dharam

OUTPUT

Your Name in Reverse Order is:
marahD

Example 3.19: Program to input a string through command line and display the number of vowels.

```
class example
{
public static void main(String[] args)
{
String str1 = args[0];
int k=0;
for(int i=0;i<str1.length();i++)
{
char s=str1.charAt(i);
switch(s)
{
    case 'A':case 'a': case 'U':case 'u':
    case 'O':case 'o': case 'E':case 'e':
    case 'I':case 'i':k++;break;
}
}
    System.out.println("Number of Vowels: "+k);
}
}
```

Execution Command (if program name is example.java):
C:\> java example Neha Rani

OUTPUT

Number of Vowels: 4

Example 3.20: Program to input a string through command line and display in the form of triangle, (for example : input→welcome)

```
w
we
wel
welc
welco
welcom
welcome
class example
{
public static void main(String[] args)
{

String str1 = args[0];

char[] arr=new char[10];
for(int i=0;i<str1.length();i++)
{
char s=str1.charAt(i);
arr[i]=s;
}
for(int i=0;i<str1.length();i++)
{
      for(int j=0;j<=i;j++)
      System.out.print(arr[j]);
      System.out.println();
}
}
}
```

Execution Command (if program name is example.java):
C:\> java example Welcome

OUTPUT

```
w
we
wel
welc
welco
welcom
welcome
```

3.3. CONSOLE INPUTS

In the previous examples you saw we can take the inputs as a command line argument, but it is not an interactive input where program must ask for input to user. Java provides some classes for this purpose.

1. Scanner class
2. BufferedReader class

3.3.1. Scanner Class

The Scanner class is the member of java.util package. To input data using Scanner class, you have to create an object of Scanner class and then used appropriate method to input data from keyboard. There are some methods of Scanner class:

Method	Description
next(), nextLine()	Used to input string from console
nextInt()	Used to input integer data from console
nextFloat()	Used to input float data from console
nextDouble()	Used to input double data from console
nextLong()	Used to input long data from console

Here are some examples to show the use of these methods.

Example 3.21: Program to input your details and display them.

```
import java.util.*;
class example
{
public static void main(String[] args)
{
Scanner sc = new Scanner(System.in);
System.out.println("Enter the Your Name:");
String name=sc.next();

System.out.println("Enter the Your Age:");
int age=sc.nextInt();

System.out.println("Enter the Your Weight in kg:");
float weight=sc.nextFloat();

System.out.println("Your Name : "+name);
System.out.println("Your Age : "+age);
System.out.println("Your Weight (kg): "+weight);
}
}
```

OUTPUT

```
Enter the Your Name:
Radhika
```

```
Enter the Your Age:
45
Enter the Your Weight in kg:
48.5
Your Name          :      Radhika
Your Age       :     45
Your Weight(kg)    :      48.5
```

Example 3.22: Program to input sides of rectangle and find the area of rectangle.

```
import java.util.*;
class example
{
public static void main(String[] args)
{
Scanner sc = new Scanner(System.in);
System.out.println("Enter the Length of Rectangle:");
int length=sc.nextInt();
System.out.println("Enter the Width of Rectangle:");
int width=sc.nextInt();
int area = length * width;
System.out.println("Area of Rectangle:"+area);
}
}
```

OUTPUT

```
Enter the Length of Rectangle:
12
Enter the Width of Rectangle:
10
Area of Rectangle: 120
```

Example 3.23: Program to input a number n find the sum of all first n natural numbers and display the factorial of n.

```
import java.util.*;
class example
{
public static void main(String[] args)
{
Scanner sc = new Scanner(System.in);

System.out.println("Enter the Number:");
int num=sc.nextInt();
int fact=1;
int sum=0;
for(int i=1; i<=num;i++)
{
```

```
fact=fact*i;
sum = sum+i;
}
System.out.println("Factorial of "+num+" is : "+fact);
System.out.println("Sum of first "+num+" natural numbers is : "+sum);

}
}
```

OUTPUT

```
Enter the Number:
5
Factorial of 5 is : 120
Sum of first 5 natural numbers is : 15
```

Example 3.24: Program to input a number and display all digits in words.

```
import java.util.*;
class example
{
public static void main(String[] args)
{
Scanner sc = new Scanner(System.in);

System.out.println("Enter the Number:");
int num=sc.nextInt();
int n=0;
int a;
while(num!=0)
{
	a= num%10;
	n= n*10+a;
	num=num/10;
}
System.out.println("Conversion in words:");
while(n!=0)
{
	a=n%10;
	n=n/10;
	switch(a)
	{
	case 0	:	System.out.print(" Zero");
			break;
	case 1	:	System.out.print(" One");
			break;
	case 2	:	System.out.print(" Two");
			break;
	case 3	:	System.out.print(" Three");
			break;
```

```
        case 4   :   System.out.print(" Four");
                     break;
        case 5   :   System.out.print(" Five");
                     break;
        case 6   :   System.out.print(" Six");
                     break;
        case 7   :   System.out.print(" Seven");
                     break;
        case 8   :   System.out.print(" Eight");
                     break;
        case 9   :   System.out.print(" Nine");
                     break;
        }
    }

    }
    }
```

OUTPUT

```
Enter the Number:
123
Conversion in words:
One Two Three
```

Example 3.25: Program to input numbers up-to the user choice and display the followings:

(*i*) Display the sum of all input numbers

(*ii*) Display the maximum number of all input numbers

(*iii*) Display the minimum number of all input numbers

(*iv*) Display how many numbers input by the user

```
import java.util.*;
class example
{
public static void main(String[] args)
{
Scanner sc = new Scanner(System.in);

int sum,max,min,count;
System.out.println("Enter a Number:");
int num=sc.nextInt();
sum=max=min=num;
count=1;
System.out.println("Do you want to continue (y/n):");
String ch= sc.next();
char choice =ch.charAt(0);
while(choice=='y' || choice=='Y')
{
```

```
System.out.println("Enter next Number:");
num=sc.nextInt();
sum = sum + num;
if(max<num) max = num;
if(min>num) min = num;
count++;
System.out.println("Do you want to continue (y/n):");
ch= sc.next();
choice =ch.charAt(0);
}
System.out.println("Total number of Inputs : "+count);
System.out.println("Sum of All Inputs       : "+sum);
System.out.println("Maximum number of Inputs : "+max);
System.out.println("Minimum number of Inputs : "+min);

}
}
```

OUTPUT

```
Enter a Number:
12
Do you want to continue (y/n):
y
Enter next Number:
2
Do you want to continue (y/n):
y
Enter next Number:
6

Do you want to continue (y/n):
n
Total number of Inputs  : 3
Sum of All Inputs        : 20
Maximum number of Inputs: 12
Minimum number of Inputs: 2
```

3.3.2. BufferedReader Class

The BufferedReader class uses some more classes to perform the input process, all these classes are the member of java.io package. When you entered data from keyboard, it uses System.in and gives the data in byte form to InputStreamReader class object, which converts the data into a character. This character is provided to BufferedReader class object to place those characters in temporary buffer area till enter key is pressed. When enter key is pressed, data can be taken from buffer using readLine() method. Whenever some exception is occur, the readLine() method throws them, all these exception called IOException. You have to handle them by exception handling concept, we will discuss them in later, but we are using for some examples.

Here are some examples to show the use of this technique.

Example 3.26: Program to input name and display with welcome greeting.

```
import java.io.*;
class example
{
	public static void main(String aa[])
	{
		String s;
		try
		{
			System.out.print("Plz enter your name : ");
			BufferedReader r = new BufferedReader(new
		InputStreamReader(System.in));
			s = r.readLine();
			System.out.println("Welcome "+s);
		} catch(Exception ex){}
	}
}
```

OUTPUT

```
Plz enter your name: Ashish Kumar
Welcome Ashish Kumar
```

Example 3.27: Program to input name and display with welcome greeting without using try and catch block.

(This is another form of previous example without using try and catch block. In this example we used throws IOException in the line of main method -see care fully.)

```
import java.io.*;
class example
{
	public static void main(String aa[]) throws IOException
	{
		String s;
			System.out.print("Plz enter your name : ");
			BufferedReader r = new BufferedReader(new
		InputStreamReader(System.in));
			s = r.readLine();
			System.out.println("Welcome "+s);

	}
}
```

OUTPUT

```
Plz enter your name: Ashish Kumar
Welcome Ashish Kumar
```

Example 3.28: Program to input two numbers and display the sum of them.

```
import java.io.*;
class example
{
	public static void main(String aa[])
	{
		int a,b,c;
		try
		{
			BufferedReader r = new BufferedReader(new
		InputStreamReader(System.in));
			System.out.print("Plz enter Ist Number : ");
			a = Integer.parseInt(r.readLine());
			System.out.print("Plz enter IInd Number : ");
			b = Integer.parseInt(r.readLine());
		}
		catch(Exception ex) {  a = 0; b = 0; }
		c = a + b;
		System.out.println("Sum is = "+c);

	}
}
```

OUTPUT

```
Plz enter Ist Number: 12
Plz enter IInd Number: 13
Sum is = 25
```

3.3.3. Some More Input Forms

Using Byte Format

You can input the data using byte format, see an example where we input a name in the form of bytes. We can enter bytes directly by using System.in class but you have to check the exception to be handle.

Example 3.29: Program to input name and display with welcome greeting.

```
class example
{
	public static void main(String aa[])
	{
		byte b[] = new byte[10];
		try
		{
			System.out.print("Plz enter your name : ");
			System.in.read(b);  // CheckedException - need to handle
			System.out.println("Welcome "+new String(b));
		}catch(Exception ex){}
```

```
    }
}
```

OUTPUT

```
Plz enter your name: Ashish Kumar
Welcome Ashish Kumar
```

Using DataInputStream Class

DataInputStream is a class of java.io package.

You can input the data using string format, see an example where we input a name in the form of string. We have used readLine() method to input string from console.

Example 3.30: Program to input name and display with welcome greeting.

```
import java.io.*;
class example
{
        public static void main(String aa[])
        {
                String s;
                try
                {
                        System.out.print("Plz enter your name : ");
                        DataInputStream d = new DataInputStream(System.in);
                        s = d.readLine();
                        System.out.println("Welcome "+s);
                }catch(Exception ex){}
        }
}
```

OUTPUT

```
Plz enter your name: Ashish Kumar
Welcome Ashish Kumar
```

3.4. MULTIPLE "MAIN()" FUNCTIONS

You can write multiple main() methods in a single file, and they will execute independently, with some simple change in execution commands. Let see an example:

Example 3.31: Program to demonstrate the multiple main() in a single file.

```
// File Name: MainExample.java
// Declaration of class First
class First
{
        public static void main(String [] args)
        {
        System.out.println("\n I am from class First");
        }
}
```

```
//Declaration of class Second
class Second
{
        public static void main(String [] args)
        {
        System.out.println("\n I am from class Second");
        }
}
//Declaration of class Third
class Third
{
        public static void main(String [] args)
        {
        System.out.println("\n I am from class Third");
        }
}
```

COMPILATION:

```
C:\> javac MainExample.java
```

EXECUTION:

```
C:\>java First
I am from class First
C:\>java Second
I am from class Second
C:\>java Third
I am from class Third
```

See carefully, what command we have given for compilation.

```
C:\> javac MainExample.java
```

The main concept behind this program is that when we compile program file with its own name, the java compiler compiles the file and makes .class files for all classes available in the file with main function. See in figure given below:

Therefore to execute the particular class we have to use the class name as follows:

```
C:\> java First

C:\>java Second

C:\>java Third
```

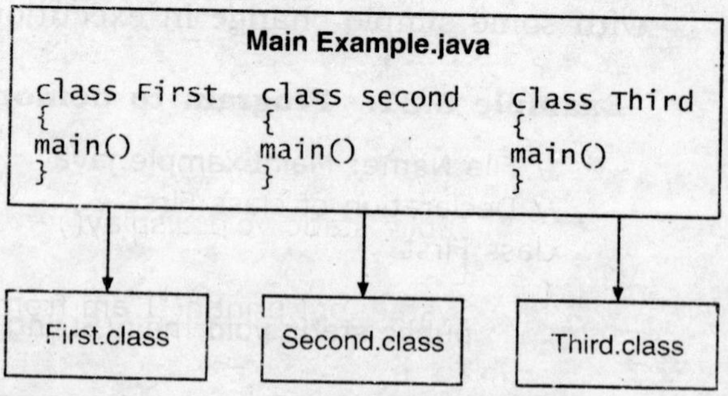

3.5. MULTI-FILE PROGRAM

In previous sections you lernt how to develop programs and how to compile and execute them. The similarity among all previous examples are that they all have their own single files. But Java also provides the facility to combine the more then one file program into single executable file. The program structure for the multi-file program is given below:

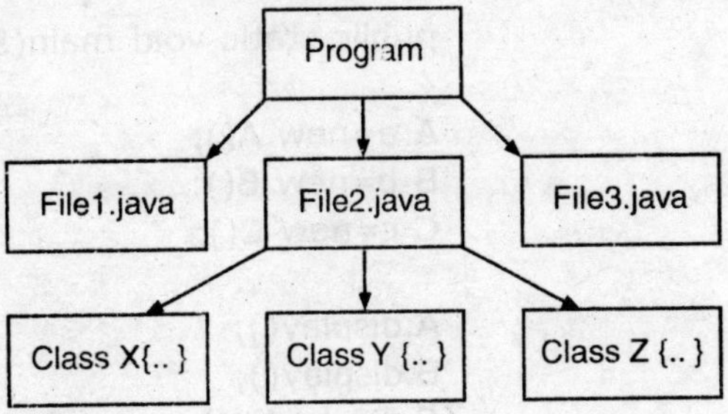

According to the above figure program may divided into several files, file may contain several classes, interfaces and etc.

These type of multi-file program needs more care at the time of compilation, they needs a special command for compilation. The syntax for the command is given below:

```
C:\> javac <filename1> <filename2>....<filenameN>
```

For Example, we have three files named prog1.java, prog2.java and prog3.java and we want to compile them into one executable file, then the command will be as follows:

```
C:\> javac prog1.cs prog2.cs prog3.cs
```

Example 3.32: To execute Program consisting four files.

```
//File Name : A.java
class A
{
        public static void display()
        {
        System.out.println("I am from class A");
        }
}

//File Name : B.java
class B
{
        public static void display()
        {
        System.out.println("I am from class B");
        }
}

//File Name : C.java
class C
{
        public static void display()
        {
        System.out.println("I am from class C");
        }
}

//File Name : Common.java
class Common
{
```

```
public static void main(String [] args)
{
A a=new A();
B b=new B();
C c=new C();

A.display();
B.display();
C.display();
System.out.println("I am from Common class");

}
}
```

COMPILATION:

```
C:\> javac  Common.java  A.java  B.java  C.java
```

EXECUTION:

```
C:\> java Common
I am from class A.
I am from class B.
I am from class C.
I am from Common class.
```

You can understand above example using the figure given below:

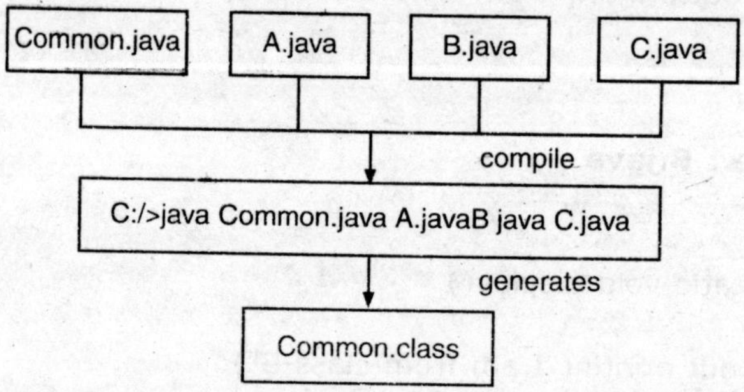

SUMMARY

- In this chapter we tried toexplain all the programming concept through programs.
- Command line argument is the technique topass the arguments parameters through the command prompt, asan input.
- We given some examplesfor comand line arguments using characters, Integers andstrings.
- Java alsosupportsconcoleinputs. In java there are sevral techniqueto input from console such as scannerclassand Buffered reader class.
- We have also given some examples, which shows-you can use multiple "main" methods in single file and multiplefiles can be execute as a single program.

EXERCISES

Q.3.1. Write a program to display your namein both smalled capital letters.

Q.3.2. Write a program to input first name,middle name and last name using command line arguments and display name using single string.

Q.3.3. Write a progra to input using scanner class from matrix (3 × 3) and display the sum of all elements.

Q.3.4. Write a program to input two numbers display in words which is large number.

Q.3.5. Write a program to display the input word in uppar case or vice verse.

Q.3.6. Write a program to display an input integer with its features like it even, prime and multiple of 1 – 9. (For example : input = 15

Features :

1. It is even number.
2. It is not prime.
3. It is multiple of 5.)

□□□

Chapter 4

Method and Classes

Whereas a subroutine represents a single task, an object can encapsulate both data (called fields in Java) and a number of different tasks or "behaviors" related to that data (called methods). Therefore objects provide another, more sophisticated type of structure that can be used to help manage the complexity of large programs. This chapter covers the creation and use of objects in Java.

CHAPTER OUTLINE

4.1. CREATING OBJECTS

When you create a class, you are creating a new data type. For Example:

```
class item
{
        int number;
        float cost;
}
```

You can use this type to declare objects of that type. The declaration of objects of a class is a two-step process.

(*i*) Declare a variable of the class type.

(*ii*) Acquire an actual, physical copy of the object and assign it to that variable.

The declaration of **item** as shown above does not define any object of **item** but only specifies *what* they will contain. Once a class has been declared, we can create variables of that type by using the class name (like any other built-in type variable). For example,

```
item x;
```

This variable **x** does not define an object. Instead it is simply a variable that can refer to an object. After that you must acquire an actual, physical copy of the object and design it to that variable. You can do this using the **new** operator. The **new** operator dynamically allocates memory for an object and returns a reference to it. For examples,

```
x = new item();
```

It is also possible to combine the two steps, declaration and memory allocation of object, into one as shown below:

```
item x = new item();
```

Here is an example to declare an object for class Exercise

```
class Exercise
{
    int a, b;
}

class Class1
{
    public static void main(String args[])
    {
    Exercise exo = new Exercise();
    }
}
```

4.2. WORKING WITH METHODS

Methods are nothing but a set of executable statements. Methods (also called as function in some programming languages like C, C++) are also the interface to the data variables of the class. Methods provide a structured approach to programming. A program can be divided into several methods.

While debugging the program, a debugger can directly jump to a particular method and make the necessary corrections. Just think of writing thousands of lines of code without any method. It would be a difficult job to debug such a program.

Advantages of Methods

- Methods are provided to give access to the data of the class. No one other than the 'methods' can directly access the data variable of class.
- A program can be divided logically.
- When the same set of code is required to be used again and again in the program, there is no need to repeat the same set of statement this set of code can be de• lared as a method, which can then be called as and when required.
- The program easy to debug

Tip: There is no support for global methods in java; every method must be defined within a class.

Method declarations

<access modifier> <static/final> <return type> <method name> <arguments list>

```
{
- - Method body;
}
```

Example:

```
public static void main(String args[])
{

}
```

Here is some example programs to understand the concept of nethods

Example 4.1: Sample program of single Method

```
public class Test
{
    public static vo'd welcome()
    {
        System.out.println("——————————————");
    }
    public static void main(String args[])
    {
    welcome();
    System.out.println("Hi-this is my first Program");
    welcome();

    }
}
```

OUTPUT:

```
——————————————
Hi-this is my first Program
——————————————
```

Example 4.2: Multiple Methods with class

```
class function
{
void first()
    {
        System.out.println("I am First Function");
    }

    void second()
    {
        System.out.println("I am Second Function");
    }
}
```

```
public class Test
{
    public static void main(String args[])
    {
        function F= new function();
        F.first();
        F.second();
        System.out.println("I am main Function");
    }
}
```

OUTPUT:

```
I am First Function
I am Second Function
I am main Function
```

Example 4.3: Multiple Classes with Methods

```
class First
{
 void first()
 {
     System.out.println("I am First's First Function");
 }

 void second()
 {
     System.out.println("I am First's Second Function");
 }
}
class Second
{
 void first()
 {
     System.out.println("I am Second's First Function");
 }

 void second()
 {
     System.out.println("I am Second's Second Function");
 }
}
public class Test
{
    public static void main(String args[])
 {
    First F = new First();
    Second S = new Second();
```

```
        F.first();
        S.first();
        F.second();
        S.second();
        System.out.println("I am main Function");
    }
}
```

OUTPUT:

```
I am First's First Function
I am Second's First Function
I am First's Second Function
I am Second's Second Function
I am main Function
```

4.3. CONSTRUCTOR

You can imagine creating a method called **initialize( )** for every class you write. The name is a hint that it should be called before using the object. Unfortunately, this means the user must remember to call the method. In Java, the class designer can guarantee initialization of every object by providing a special method called a *constructor.*

> **Concept**
>
> *Constructor is a method of class to allocate the memory of object and initilize the variable of class*

C++ introduced the concept of a *constructor*, a special method automatically called when an object is created. Java also adopted the constructor, and in addition has a garbage collector that automatically releases memory resources when they're no longer being used.

If a class has a constructor, Java automatically calls that constructor when an object is created, before users can even get their hands on it. The name of the constructor is the same as the name of the class. It makes sense that such a method will be called automatically on initialization.

Characteristics of constructors

- Constructors do not have return type or value (not even void).
- They have same name as class name.
- Constructors cannot be inherited, though a derived class can call the base class constructors.
- Java automatically calls the constructor when an object is created.

Here's a simple class with a constructor:

Example 4.4: Demonstration of a simple constructor

```
class Rock
{
 // This is the constructor
 Rock()
```

```
 {
 System.out.println("Creating Rock");
 }
}

public class Test
{
 public static void main(String[] args)
 {
     for(int i = 0; i < 10; i++)
     new Rock();
 }
}
```

OUTPUT:

```
Creating Rock
Creating Rock
Creating Rock
Creating Rock
Creating Rock
Creating Rock
Creating Rock
Creating Rock
Creating Rock
Creating Rock
```

Now, when an object is created:

```
new Rock();
```

storage is allocated and the constructor is called. It is guaranteed that the object will be properly initialized before you can get your hands on it.

Example 4.5: Constructor with a parameter

```
class Rock
{
 Rock(int i)
 {
     System.out.println("Creating Rock number " + i);
 }
}
public class Test
{
 public static void main(String[] args)
 {
     for(int i = 0; i < 10; i++)
     new Rock(i);
 }
}
```

OUTPUT:

```
Creating Rock number 0
Creating Rock number 1
Creating Rock number 2
Creating Rock number 3
Creating Rock number 4
Creating Rock number 5
Creating Rock number 6
Creating Rock number 7
Creating Rock number 8
Creating Rock number 9
```

4.4. GARBAGE COLLECTION

Many other object-oriented languages require that you keep track of all of the objects you create and that you explicitly destroy them when they are no longer needed. This technique of managing memory is tedious and often error-prone. Java allows you to create as many objects as you want (limited, of course, by what your system can handle), and you never have to worry about destroying them. The Java runtime environment deletes objects when it determines that they are no longer being used. This process is called *garbage collection.*

> **Concept**
>
> *The garbage collection is the process, in which Java runtime environment destroy the objects when it determines that they are no longer needed.*

An object is eligible for garbage collection when there are no more references to that object. References that are held in a variable are naturally dropped when the variable goes out of scope. Or you can explicitly drop an object reference by setting the variable to null.

4.4.1. The Garbage Collector

The Java platform has a garbage collector that periodically frees the memory used by objects that are no longer needed. The Java garbage collector is a mark-sweep garbage collector. A mark-sweep garbage collector scans dynamic memory areas for objects and marks those that are referenced. After all possible paths to objects are investigated, unmarked objects (unreferenced objects) are known to be garbage and are collected. (A more complete description of Java's garbage collection algorithm might be "a compacting, mark-sweep collector with some conservative scanning.")

The garbage collector runs in a low-priority thread and runs either *synchronously* or *asynchronously* depending on the situation and the system on which Java is running. It runs synchronously when the system runs out of memory or in response to a request from a Java program.

The Java garbage collector runs *asynchronously* when the system is idle, but it does so only on systems, such as Windows 95/NT, that allow the Java runtime environment to note when a thread has begun and to interrupt another thread. As soon as another thread becomes active, the garbage collector is asked to get to a consistent state and terminate.

4.4.2. Finalization

Before an object gets garbage-collected, the garbage collector gives the object an opportunity to clean up after itself through a call to the object's finalize method. This process is known as *finalization.*

Most programmers don't have to worry about implementing finalize. In rare cases, however, a programmer might have to implement a finalize method to release resources, such as native peers, that aren't under the control of the garbage collector.

The finalize method is a member of the Object class, which is the top of Java's class hierarchy and the parent of everything. A class can override the finalize method to perform any finalization necessary for objects of that type. If you override finalize, then your implementation of the method should call super.finalize as the last thing it does. Overriding Methods discussed in next chapter.

4.5. METHOD OVERLOADING

Another powerful object-oriented technique is method overloading. Method overloading enables you to specify different types of information (parameters) to send to a method. To overload a method, you declare another version with the same name but different parameter.

For example, the move method for the Alien class could have two different versions: one general movement and one for moving to a specific location. The general version moves the alien based on its current state. The declaration for this version follows:

> **Concept**
>
> *Multiple methods of same name with different arguments (parameters) in a single class is called method overloading.*

```
void move()
{
        // move the alien
}
```

To enable the alien to move to a specific location, you overload the move method with a version that takes x and y parameters, which specify the location to move. The overloaded version of move follows:

```
void move(int x, int y)
{
        // move the alien to position x,y
}
```

Notice that the only difference between the two methods is the parameter lists; the first move takes no parameters while the second move takes two integers.

You may be wondering how the compiler knows which method is being called in a program, when they both have the same name. The compiler keeps up with the parameters for each method along with the name. When a call to a method is encountered in a program, the compiler checks the name and the parameters to determine which overloaded method is being called. In this case, calls to the move methods are easily distinguishable by the absence or presence of the integer parameters.

Example 4.6: Method Overloading of an area method

```
class Shape
{
        float area(float r)
        {
                return 4.14*r*r;
        }
        float area(float l, float b)
        {
                return (l*b)/2;
        }
                int area(int a)
        {
                return a*a;
        }

}
public class Test
{
        public static void main(String[] args)
        {
                Shape A = new Shape();
                float circle = A.area(7.2f);
                System.out.println("Area of circle is :"+ circle);

                float rect = A.area(2.0f,8.4f);
                System.out.println("Area of Rectangle is :"+ rect);

                int square = A.area(2);
                System.out.println("Area of Square is :"+ square);

        }
}
```

OUTPUT:

```
Area of circle is : 162.7776
Area of Rectangle is : 8.4
Area of square is : 4
```

4.5.1. Overloading Constructor

In Java (and C++), another factor forces the overloading of method names: the constructor. Because the constructor's name is predetermined by the name of the class, there can be only one constructor name. But what if you want to create an object in more than one way? For example, suppose you build a class that can initialize itself in a standard way or by reading information from a file. You need two constructors, one that takes no arguments (the *default* constructor), and one that takes a **String** as an argument, which is the name of the file from which to initialize the object. Both are constructors, so they must have the same name—the name of the class. Thus, *method overloading* is essential to allow the same method name to be used with different argument types. And although method overloading is a must for constructors, it's a general convenience and can be used with any method.

Here's an example that shows both overloaded constructors and overloaded ordinary methods:

Example 4.7: Method and constructor Overloading

```
class TIME
{
    int H,M,S;
    TIME()
    {
        H = 0;
        M = 0;
        S = 0;
}
TIME(int h, int m)
{
        H = h;
        M = m;
        S = 0;
}
TIME(int h, int m, int s)
{
    H = h;
    M = m;
    S = s;
}
void add(TIME T)
{
    H = H + T.H;
    M = M + T.M;
    S = S + T.S;
}
void add(TIME Z, TIME T)
{
    H = Z.H + T.H;
    M = Z.M + T.M;
    S = Z.S + T.S;
}
void show()
{
    System.out.println( H + ":" + M + ":" + S );
}
}
public class TimeExample
```

```
{
 public static void main(String[] args)
 {
     TIME A = new TIME();
     TIME B = new TIME(2,30);
     TIME C = new TIME(4,50,35);

     System.out.println("The value of Time Object A:");
     A.show();

     System.out.println("The value of Time Object B:");
     B.show();

     System.out.println("The value of Time Object C:");
     C.show();

     System.out.println("The Sum of Time Object B & C through Object
     A:"); A.add(B,C); A.show();

     System.out.println("The Sum of Time Object B & C through Object
     B:"); B.add(C); B.show();
 }
}
```

OUTPUT:

```
The value of Time Object A:
0 : 0 : 0
The value of Time Object B:
2 : 30 : 0
The value of Time Object C:
4 : 50 : 35
The Sum of Time Object B & C through Object A:
 7 : 20 : 35
The Sum of Time Object B & C through Object B:
7 : 20 : 35}
```

4.6. ARRAYS

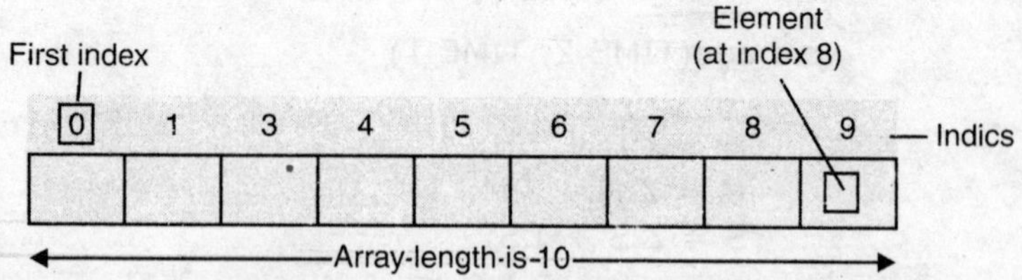

Fig. 4.1. An array of ten elements

An *array* is a container object that holds a fixed number of values of a single type. The length of an array is established when the array is created. After creation, its length is fixed. You've seen an example of arrays already, in the main method of the "Hello World!" application. This section discusses arrays in greater detail.

Each item in an array is called an *element*, and each element is accessed by its numerical *index*. As shown in the above illustration, numbering begins with 0. The 9th element, for example, would therefore be accessed at index 8.

The following program example 4.8 , creates an array of integers, puts some values in it, and prints each value to standard output.

Example 4.8: Array Demo

```
class ArrayDemo {
        public static void main(String[] args) {
        int[] anArray; // declares an array of integers

        anArray = new int[10]; // allocates memory for 10 integers

        anArray[0] = 100; // initialize first element
        anArray[1] = 200; // initialize second element
        anArray[2] = 300; // etc.
        anArray[3] = 400;
        anArray[4] = 500;
        anArray[5] = 600;
        anArray[6] = 700;
        anArray[7] = 800;
        anArray[8] = 900;
        anArray[9] = 1000;

        System.out.println("Element at index 0: " + anArray[0]);
        System.out.println("Element at index 1: " + anArray[1]);
        System.out.println("Element at index 2: " + anArray[2]);
        System.out.println("Element at index 3: " + anArray[3]);
        System.out.println("Element at index 4: " + anArray[4]);
        System.out.println("Element at index 5: " + anArray[5]);
        System.out.println("Element at index 6: " + anArray[6]);
        System.out.println("Element at index 7: " + anArray[7]);
        System.out.println("Element at index 8: " + anArray[8]);
        System.out.println("Element at index 9: " + anArray[9]);
    }
}
```

OUTPUT:

```
Element at index 0: 100
Element at index 1: 200
Element at index 2: 300
Element at index 3: 400
Element at index 4: 500
Element at index 5: 600
Element at index 6: 700
Element at index 7: 800
Element at index 8: 900
Element at index 9: 1000
```

In a real-world programming situation, you'd probably use one of the supported *looping constructs* to iterate through each element of the array, rather than write each line individually as shown above. However, this example clearly illustrates the array syntax. You'll learn about the various looping constructs (for, while, and do-while) in the Control statement section.

4.6.1. Declaring a Variable to Refer to an Array

The above program declares anArray with the following line of code:

```
int[] anArray;    // declares an array of integers
```

Like declarations for variables of other types, an array declaration has two components: the array's type and the array's name. An array's type is written as *type*[], where *type* is the data type of the contained elements; the square brackets are special symbols indicating that this variable holds an array. The size of the array is not part of its type (which is why the brackets are empty). An array's name can be anything you want, provided that it follows the rules and conventions as previously discussed in the identifiers section. As with variables of other types, the declaration does not actually create an array -- it simply tells the compiler that this variable will hold an array of the specified type.

Similarly, you can declare arrays of other types:

```
byte[] anArrayOfBytes;
short[] anArrayOfShorts;
long[] anArrayOfLongs;
float[] anArrayOfFloats;
double[] anArrayOfDoubles;
boolean[] anArrayOfBooleans;
char[] anArrayOfChars;
String[] anArrayOfStrings;
```

You can also place the square brackets after the array's name:

```
float anArrayOfFloats[]; // this form is discouraged
```

However, convention discourages this form; the brackets identify the array type and should appear with the type designation.

4.6.2. Creating, Initializing, and Accessing an Array

One way to create an array is with the new operator. The next statement in Example 4.8 the ArrayDemo program allocates an array with enough memory for ten integer elements and assigns the array to the anArray variable

```
anArray = new int[10]; // create an array of integers
```

If this statement were missing, the compiler would print an error like the following, and compilation would fail:

```
ArrayDemo.java:4: Variable anArray may not have been initialized.
```

The next few lines assign values to each element of the array:

```
anArray[0] = 100; // initialize first element
anArray[1] = 200; // initialize second element
anArray[2] = 300; // etc.
```

Each array element is accessed by its numerical index:

```
System.out.println("Element 1 at index 0: " + anArray[0]);
System.out.println("Element 2 at index 1: " + anArray[1]);
System.out.println("Element 3 at index 2: " + anArray[2]);
```

Alternatively, you can use the shortcut syntax to create and initialize an array:

```
int[] anArray = {100, 200, 300, 400, 500, 600, 700, 800, 900, 1000};
```

Here the length of the array is determined by the number of values provided between { and }.

You can also declare an array of arrays (also known as a *multidimensional* array) by using two or more sets of square brackets, such as String[][] names. Each element, therefore, must be accessed by a corresponding number of index values.

In the Java programming language, a multidimensional array is simply an array who components are themselves arrays. This is unlike arrays in C or Fortran. A consequence of this is that the rows are allowed to vary in length, as shown in the following example 4.9 :

Example 4.9: Multi Dimensional Array

```
class MultiDimArrayDemo
{
    public static void main(String[] args)
    {
    String[][] names = {{"Mr. ", "Mrs. ", "Ms. "},
                        {"Smith", "Jones"}};

    System.out.println(names[0][0] + names[1][0]); //Mr. Smith
    System.out.println(names[0][2] + names[1][1]); //Ms. Jones
    }
}
```

OUTPUT:

```
Mr. Smith
Ms. Jones
```

Finally, you can use the built-in length property to determine the size of an array. The code

```
System.out.println(anArray.length);
```

will print the array's size to standard output.

4.6.3. Copying Arrays

The System class has an arraycopy method that you can use to efficiently copy data from one array into another:

```
public static void arraycopy(Object src, int srcPos, Object dest, int destPos, int length)
```

The two Object arguments specify the array to copy *from* and the array to copy *to*. The three int arguments specify the starting position in the source array, the starting position in the destination array, and the number of array elements to copy.

The following program, example 4.10, declares an array of char elements, spelling the word "decaffeinated". It uses arraycopy to copy a subsequence of array components into a second array:

Example 4.10: Array Copy Demo

```
class ArrayCopyDemo
{
public static void main(String[] args)
        {
        char[] copyFrom = { 'd', 'e', 'c', 'a', 'f', 'f', 'e',
        'i', 'n', 'a', 't', 'e', 'd' };
        char[] copyTo = new char[7];
        System.arraycopy(copyFrom, 2, copyTo, 0, 7);
        System.out.println(new String(copyTo));
        }
}
```

OUTPUT:

```
caffein
```

4.6.4. Example Programs for Arrays.

Example 4.11: Pascal Triangle using dynamic array:

```
public class Gurdy
{
public static void main(String [] args)
{
        int [] current = null;
        int [] previous = {};
        for (int r=1; r<17; r++) {
        current = new int[previous.length+1];
        current[0] = 1;
        current[previous.length] = 1;
        for (int c=1; c<previous.length; c++) {
                current[c] = previous[c]+previous[c-1];
                }
        previous = current;
        }

    for (int r=0; r<current.length; r++) {
        System.out.print(" "+current[r]);
        }
    System.out.println("");
    }
}
```

Example 4.12: Pascal Triangle using Two Dimensional array

```
public class Hurdy
{
public static void main(String [] args)
{

        int [][] tri = new int[20][];

        for (int r=0; r<tri.length; r++)
        {
        tri[r] = new int[r+1];
        tri[r][0] = 1;
        tri[r][r] = 1;
        for (int c=1; c<r; c++)
        {
                tri[r][c] = tri[r-1][c]+tri[r-1][c-1];
        }
 }
 for (int r=0; r<tri.length; r++)
 {
        for (int c=0; c<tri[r].length; c++)
        {
                System.out.print(" "+tri[r][c]);

        }
        System.out.println("");

  }
 }
 }
```

4.7. STRING HANDLING

Strings in Java are *not* null-terminated arrays of characters as they are in C. Instead, they are instances of the java.lang.String class.

Internally, a String encapsulates an array of characters. The reason for this encapsulation is that the String object ensures that the string will be maintained properly. It also encapsulates a number of useful methods that can be performed on a string. The most important thing to remember is that a String is **immutable**, which means that after you assigned your String object a value, it cannot change anymore during its lifetime. If you want to manipulate the characters, consider using the StringBuffer class.

> **Concept**
>
> *String is a class, not a data type in Java.*

4.7.1. String Creation

There are two ways to create a String object:

1. Implicitly
2. Explicitly

1. **Implicitly:** when you use a string literal, like "Hello, world!", Java automatically creates a String object for you.

For example:

```
String s = "Hello, world!";
```

- You can use all String methods on a string literal, example :
  s = "Hello, World!".substring(7); // assigns "world!" to s

2. **Explicitly:** when you use the new operator to instantiate a String object.

For example:

```
String s = new String("Hello, world!");
```

- This will call the String class constructor that takes another String as parameter. (remember, "Hello, world!" is represented internally as a String object!).The String object has other constructors to create Strings as shown in the following test program:

```
public class TestProg
{
public static void main(String args[])
{
	try {
		String s;
		StringBuffer sb = new StringBuffer("A StringBuffer");
		// create a char array
		char ca[] = "A character array!".toCharArray();
		// create a byte array containing the default encoding
		// ISO-Latin-1
		byte ba[] = "A byte array!".getBytes();
		// create a byte array containing Unicode
		byte bau[] = "A Unicode byte array!".getBytes("Unicode");

		// create an empty String
		s = new String();
		System.out.println(s); // prints out nothing

		// create a String from a byte array
		s = new String(ba);
		System.out.println(s); // prints out "A byte array!"

		// create a String from a byte array from index 2 (inclusive),
		// take 1 byte
		s = new String(ba, 2, 1);
		System.out.println(s); // prints out "byte"
```

```
            // create a String from a byte subarray containing Unicode
            s = new String(bau, "Unicode");
            System.out.println(s); // prints out "A Unicode byte array!"

            // create a String from a byte subarray containing Unicode
            s = new String(bau, 0, 20, "Unicode");
            System.out.println(s); // prints out "A Unicode"

            // create a String from a StringBuffer
            s = new String(sb);
            System.out.println(s);    //    prints    out    "A    StringBuffer"

            // create a String from a character subarray
            s = new String(ca, 2, 4);
            System.out.println(s); // prints out "char"

            // create a String from a character array
            s = new String(ca);
            System.out.println(s); // prints out "A character array!"
        }
        catch(java.io.UnsupportedEncodingException e) {
            System.out.println(e);
        }
    }
}
```

- Notice that you need to catch java.io.UnsupportedEncodingException for the methods in which you have to specify an encoding.

4.7.2. Important String Operations

Java has several string operations because strings are a common and important part of programming. These operation includes :

String Length

Use the length() method from the String class. The following example is self-explanatory:

```
String s = "hello, world!";
System.out.println(s.length()); // prints out 13
```

Character Array to String conversion

The String class has a useful constructor to convert an array to String, as you can see in the following example:

```
public class TestProg
{
    public static void main(String args[])
    {
    char [] array = {'H','e','l','l','o',',',' ','w','o','r','l','d','!' };
    String s = new String(array);
```

```
        System.out.println(s); // prints out: Hello, world!
        }

}
```

String to Character Array conversion

And you can convert a string to character array also, the String class has a useful method doing just that:

```
public class TestProg
{
    public static void main(String args[])
    {
        String s = "Hello, world!";
        char array[] = s.toCharArray();

        for (int i=0; i<array.length; i++)
        System.out.print(array[i]);
        System.out.println();

    }

}
```

OUTPUT:

```
Hello, world!
```

String to lowercase or uppercase

The following example is self-explanatory:

```
public class TestProg
{
    public static void main(String args[])
    {
        String s = "Hello, World!";
        String supper = s.toUpperCase();
        String slower = s.toLowerCase();

        System.out.println(supper); // prints out: HELLO, WORLD!
        System.out.println(slower); // prints out: hello, world!
    }
}
```

Note: for internationalization purposes, the methods *toUpperCase()* and *toLowerCase()* perform some checking for languages as German and Turkish to do the correct conversion. If you need these functions in time-critical applications, you might want to consider rewriting them without these checking.

Concatenate two Strings

There are two ways to concatenate one string to another. The most common is to use the (rather ambiguous) + operator:

```
String s1 = "Hello, ";
String s2 = "world!";
String s3 = s1 + s2;
System.out.println(s3); // prints out: Hello, world!
```

The String class also provides a method *concat* that does just the same:

```
String s1 = "Hello, ";
String s2 = "world!";
String s3 = s1.concat(s2);
System.out.println(s3); // prints out: Hello, world!
```

In the previous example, the method concat is invoked on s1. s1 remains unchanged and a completely new String will be created. A String is immutable, Once you assigned it a value, it cannot change anymore during its lifetime. You will always have to create a new String object.

Comparison of two Strings

Use the method java.lang.String.equals(Object). Will return true if it matches and false if it doesn't.

```
String a = "some text";
String b = "some text";

if (a.equals(b)) {
      System.out.println("String a is equal to String b");
} else {
      System.out.println("String a is NOT equal to String b");
}
```

public int compareTo(String str)

where, str is the String being compared to the invoking String. The compareTo() method returns an int value as the result of String comparison. If the value is less than zero, then the invoking string is less than the argument string. If value is zero means the invoking string and the argument string are same. If the value is greater than zero it means the invoking string is greater than the argument string. The String class also has the compareToIgnoreCase() method that compares two strings without taking into consideration their case difference.

4.7.3. All String Classes

String Class:

The String class represents character strings. All string literals in Java programs, such as "abc", are implemented as instances of this class.

Strings are constant; their values cannot be changed after they are created. String buffers support mutable strings. Because String objects are immutable they can be shared. For example:

```
String str = "abc";
```

is equivalent to:

```
char data[] = {'a', 'b', 'c'};
String str = new String(data);
```

Here are some more examples of how strings can be used:

```
System.out.printl("abc");
String cde = "cde";
System.out.println("abc" + cde);
String c = "abc".substring(2,3);
String d = cde.substring(1, 2);
```

The class String includes methods for examining individual characters of the sequence, for comparing strings, for searching strings, for extracting substrings, and for creating a copy of a string with all characters translated to uppercase or to lowercase.

The Java language provides special support for the string concatenation operator (+), and for conversion of other objects to strings. String concatenation is implemented through the StringBuffer class and its append method. String conversions are implemented through the method toString, defined by Object and inherited by all classes in Java.

StringBuffer Class:

The StringBuffer class is a thread-safe, mutable sequence of characters. A string buffer is like a String , but can be modified. At any point in time it contains some particular sequence of characters, but the length and content of the sequence can be changed through certain method calls.

String buffers are safe for use by multiple threads. The methods are synchronized where necessary so that all the operations on any particular instance behave as if they occur in some serial order that is consistent with the order of the method calls made by each of the individual threads involved.

The principal operations on a StringBuffer are the append and insert methods, which are overloaded so as to accept data of any type. Each effectively converts a given datum to a string and then appends or inserts the characters of that string to the string buffer. The append method always adds these characters at the end of the buffer; the insert method adds the characters at a specified point.

For example, if z refers to a string buffer object whose current contents are "start", then the method call z.append("le") would cause the string buffer to contain "startle", whereas z.insert(4, "le") would alter the string buffer to contain "starlet".

In general, if sb refers to an instance of a StringBuffer, then sb.append(x) has the same effect as sb.insert(sb.length(), x).

Whenever an operation occurs involving a source sequence (such as appending or inserting from a source sequence) this class synchronizes only on the string buffer performing the operation, not on the source.

Every string buffer has a capacity. As long as the length of the character sequence contained in the string buffer does not exceed the capacity, it is not necessary to allocate a new internal buffer array. If the internal buffer overflows, it is automatically made larger.

StringBuilder Class:

The StringBuilder class is mutable sequence of characters. This class provides an API compatible with StringBuffer, but with no guarantee of synchronization. This class is designed for use as a drop-in replacement for StringBuffer in places where the string buffer was being used by a single thread (as is generally the case). Where possible, it is recommended that this class be used in preference to StringBuffer as it will be faster under most implementations.

The principal operations on a StringBuilder are the append and insert methods, which are overloaded so as to accept data of any type. Each effectively converts a given datum to a string and then appends or inserts the characters of that string to the string builder. The append method always adds these characters at the end of the builder; the insert method adds the characters at a specified point.

For example, if z refers to a string builder object whose current contents are "start", then the method call z.append("le") would cause the string builder to contain "startle", whereas z.insert(4, "le") would alter the string builder to contain "starlet".

In general, if sb refers to an instance of a StringBuilder, then sb.append(x) has the same effect as sb.insert(sb.length(), x). Every string builder has a capacity. As long as the length of the character sequence contained in the string builder does not exceed the capacity, it is not necessary to allocate a new internal buffer. If the internal buffer overflows, it is automatically made larger.

Differences:

String is immutable whereas StringBuffer and StringBuilder can change their values.

The only difference between StringBuffer and StringBuilder is that StringBuilder is unsynchronized whereas StringBuffer is synchronized. So when the application needs to be run only in a single thread then it is better to use StringBuilder. StringBuilder is more efficient than StringBuffer.

Criteria to choose among String, StringBuffer and StringBuilder

- If your text is not going to change use a String Class because a String object is immutable.
- If your text can change and will only be accessed from a single thread, use a StringBuilder because StringBuilder is unsynchronized.
- If your text can changes, and will be accessed from multiple threads, use a StringBuffer because StringBuffer is synchronous

SUMMARY

In this chapter is we have explained how to deal with methods and objects.

— In java there is no support for global methods, every method must be defined within a class.

— You already knows that methods are simple operations which are associated with objects or class. The methods are provides some useful functionality.

— Methods may be return type of void, method may take arguments or without arguments.

— Method overloading is a concept of overloading a method with same name and different arguments.

— Java has given a new concept of garbage collection, which is the process of releasing the memory occupied by the objects that are no longer needed. This is an automatic process.

— We have also explained the concept of Arrays, how to create, how to initilize and how to access the elements of array in the program.

SOLVED PROBLEMS

Q.4.1. What is an object in Java?

Ans. In Java the Object class is the ultimate superclass of every other object type. All objects are extended from the Object class, either directly or by inheritance through any number of parent classes. If a class does not explicitly extend any named class, it implicitly extends the Object class. An object with a small o is the word used to describe an instance of a Java class.

Q.4.2. What's the difference between an object and a class?

Ans. An class merely defines what an object will be if it is created. An object is an actual, created instance of a class. For example, A class is like a blueprint for a house, while the object is the actual house. A single blueprint can be used to make many different houses.

Q.4.3. What's the difference between a local variable and a member variable?

Ans. A local variable is only accessible within the method where it's declared. A member variable, a field, is accessible in any method in the class.

Q.4.4. Can objects be used in place of arrays?

Ans. Yes, sometimes it is sensible to use an object to carry other object references instead of an array. For instance, you could issue an object as the return value of a method that must return multiple object references.

Q.4.5. Can we use the same variable name in two methods?

Ans. It is possible to use the same variable names for method local variables in two separate methods. However, all variables in a single method must have different names.

```
public class MethodLocalVariables
{
    void testOne()
    {
        int test;
    }
    void testTwo()
    {
    int test;
    }
}
```

Q.4.6. Explain Garbage collection mechanism in Java.

Ans. Garbage collection is a complex mechanism, but to put it in a simple way, it is the process of identifying objects which are no longer referenced by the program and discarded .The finalize() methods of these objects are called and resources occupied by the objects are released to the system. Garbage collection is carried on a separate thread and runs continuously.

Q.4.7. What do you know about the garbage collector?

Ans. Garbage collector is used to recollect memory from the unused objects. They are the objects that are no longer needed because of function or class scope is going to finish.

Q.4.8. Can the main method be overloaded?

Ans. Yes, any Java method can be overloaded, provided there is no final method with the same signature already. The Java interpreter will only invoke the standard entry point signature for the main method, with a string array argument, but your application can call its own main method as required.

Q.4.9. What's the index of the first component of a one hundred component array? of a two hundred component array? of a one thousand and twenty-four component array?

Ans. The index of the first component of an array is always 0.

Q.4.10. What's the index of the last component of a one hundred component array? of a two hundred component array? of a one thousand and twenty-four component array?

Ans. 99, 199, and 1023 respectively. The index of the last component of an array is always one less than the length of the array.

Q.4.11. What's the mininum number of components an array can have?

Ans. An array can have as few as zero components; that you can declare an array with no components at all. However you cannot declare an array with a negative number of components. If you try, a java.lang.NegativeArraySizeException will be thrown at runtime.

Q.4.12. What's the difference between a brace and a bracket?

Ans. { or } is a brace. [or] is a bracket. Braces are used to initialize arrays. Brackets are used to select a component of an array by specifying its index.

Q.4.13. Write a program that randomly fills a 3 by 4 by 6 array, then prints the largest and smallest values in the array.

Ans.

```
class array346 {

public static void main(String[] args) {

double[][][] array = new double[3][4][6];

 for (int i = 0; i < 3; i++) {
      for (int j = 0; j < 4; j++) {
      for (int k = 0; k < 6; k++) {
      array[i][j][k] = Math.random();
      }
 }
}

double max = array[0][0][0];
double min = array[0][0][0];

for (int i = 0; i < 3; i++) {
     for (int j = 0; j < 4; j++) {
     for (int k = 0; k < 6; k++) {
          if (max < array[i][j][k]) max = array[i][j][k];
          if (min > array[i][j][k]) min = array[i][j][k];
          }
     }
}
System.out.println("Maximum: " + max);
System.out.println("Minimum: " + min);

}
```

```
}
```

Q.4.14. Write a Complex class that represents a single complex number and includes methods for all the usual operations, i.e. addition, subtraction, multiplication, division, etc.

Ans:

```
public class Complex {

    private double u;
    private double v;

    Complex (double x, double y) {

    u=x;
    v=y;

  }

  public double Real () {

    return u;

  }

  public double Imaginary () {

  return v;

  }

public double Magnitude () {

  return Math.sqrt(u*u + v*v);

  }

  public double Arg () {

  // Need to watch out for zero
  return Math.atan(v/u);

  }

  // Add z to w; i.e. w += z
  public Complex Plus (Complex z) {
  return new Complex(u + z.u, v + z.v);

  }
```

```
// Subtract z from w
public Complex Minus (Complex z) {

return new Complex(u - z.u, v - z.v);

}

public Complex Times (Complex z) {

return new Complex(u*z.u - v*z.v, u*z.v + v*z.u);

}
// divide w by z
public Complex DivideBy (Complex z) {

double rz = z.Magnitude();

return new Complex((u * z.u + v * z.v)/(rz*rz),(v * z.u - u * z.v)
/(rz*rz));

}

}
```

Notice especially that x and y are now private. They cannot be accessed by external code even if we want them to be.

EXERCISES

Q.4.1. What do you mean by invocation of methods?

Q.4.2. Differentiate between actual and formal parameters?

Q.4.3. What are the advantages of methods?

Q.4.4. What is the scope of any variable?

Q.4.5. What is concept of constructor? Why it is needed?

Q.4.6. Can we use constructors with parameters?

Q.4.7. What do you mean by method overloading? Is it possible with constructors?

Q.4.8. Explain the concept of Garbage collection. What is the alternative to garbage collection?

Q.4.9. What is the difference between garbage collector and finalization?

Q.4.10. What is meant by the terms subclass and superclass?

Q.4.11. What does the computer do when it executes the following statement? Try to give as complete an answer as possible.
color[ ] palette = new color [12];

Q.4.12. What does it mean to sort an array?

Q.4.13. What is meant by basetype of an array?

Q.4.14. What is meant by a dynamic array? What is the advantage of a dynamic array over a regular array?

Q.4.15. Show the exact output produced by the following code segment:

```
char [ ] [ ] pic = new char [6] [6];
for (int i = 0; i < 6; i++)
for (int j = 0; j < 6; j ++)
{
        if (i = = j || i = = 0 || i = = 5)
                pic [i] [j] = '*';
        else
                pic [i] [j] = 'a'
}
        for (int i = 0; i < 6; i++)
        {
        for (int j = 0; j < 6; j ++)
        System.out.print (pic [i][j]);
        System.out.println();
}
```

Q.4.16. Write a program that finds the largest value in an array of integers.

Q.4.17. Suppose that temperature measurements were made on each day of 1999 in each of 100 cities. The measurements have been stored in an array

```
int [ ] [] temps = new int [100] [365];
```

where temps [c][d] holds the measurement for city number c on the dth day of the year. Write a code segment that will print out the average temperature, over the course of the whole year, for each city. The average temperature for a city can be obtained by adding up all 365 measurements for that city and dividing the answer by 365.0.

Q.4.18. Suppose that a class, Employee is defined as follows:

```
class Employee
{
        String lastName;
        String FirstName:
        double hourlywage:
        int yearswithCompany;
}
```

Suppose that data about 100 employees is already stored in an array.

```
        Employee [ ] employeeData = new Employee [100];
```

Write a code segment that will output the firstname, lastname, and hourly wage of each employee who has been with the company for 20 years or more.

Q.4.19. Suppose that A has been declared and initilized with the statement:

```
        double [ ] A = new double [20];
```

and suppose that A has already been filled with 20 values. Write a program segment that will find the average of all the non-zero numbers in the array. Declare any variables that you needed.

Chapter 5

Inheritance, Packages and Interfaces

5.1. INHERITANCE

Inheritance means taking an existing class and adding functionality by deriving a new class from it.

Object-oriented programming allows classes to *inherit* commonly used state and behavior from other classes.

A class that is derived from another class is called a *subclass* (also a *derived class, extended class*, or *child class*). The class from which the subclass is derived is called a *superclass* (also a *base class* or a *parent class*).

Excepting Object, which has no superclass, every class has one and only one direct superclass (single inheritance). In the absence of any other explicit superclass, every class is implicitly a subclass of Object.

Classes can be derived from classes that are derived from classes that are derived from classes, and so on, and ultimately derived from the topmost class, Object. Such a class is said to be *descended* from all the classes in the inheritance chain stretching back to Object.

The idea of inheritance is simple but powerful: When you want to create a new class and there is already a class that includes some of the code that you want, you can derive your new class from the existing class. In doing this, you can reuse

CHAPTER OUTLINE

the fields and methods of the existing class without having to write (and debug!) them yourself.

> **Concept**
>
> *Inheritance is a mechanism of sharing the members (member data and member function) among the classes.*

A subclass inherits all the *members* (fields, methods, and nested classes) from its superclass. Constructors are not members, so they are not inherited by subclasses, but the constructor of the superclass can be invoked from the subclass.

5.1.1. Defining subclass

In this example, Bicycle now becomes the *superclass* of MountainBike, RoadBike, and TandemBike. In the Java programming language, each class is allowed to have one direct superclass, and each superclass has the potential for an unlimited number of *subclasses*:

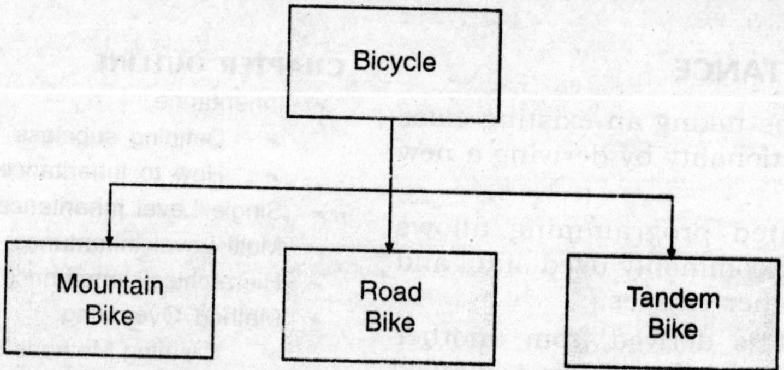

Fig. 5.1. A hierarchy of bicycle classes

The syntax for creating a subclass is simple. At the beginning of your class declaration, use the *extends* keyword, followed by the name of the class to inherit from:

```
class MountainBike extends Bicycle
{
// new fields and methods defining a mountain bike would go here
}
```

This gives MountainBike all the same fields and methods as Bicycle, yet allows its code to focus exclusively on the features that make it unique. This makes code for your subclasses easy to read. However, you must take care to properly document the state and behavior that each superclass defines, since that code will not appear in the source file of each subclass.

5.1.2. How to use inheritance

We will first create a parent class called Person. We will then create a child class called Student. Then we will create a program to use the Student class. The following is a Person class which has a variable for the person's name. There are also set and get methods for the name.

```
class Person
{
    private String name;
```

```
    public void setName(String n)
    {
        name = n;
    }
    public String getName()
    {
    return name;
    }
}
```

Now we will create the Student class that has a variable for the student number and the get and set methods for student number. The extends keyword is used to inherit from the Person class in the following example.

```
class Student extends Person
    {
    private String stuNum;

    public void setStuNum(String sn)
    {
        stuNum = sn;
    }

    public String getStuNum()
    {
        return stuNum;
    }
}
```

Finally we have to create a program that will instantiate a Student object, set the name and student number and then print the values using the get methods.

```
public class TestInheritance
{
    public static void main(String[] args)
    {
        Student stu = new Student();
        stu.setName("John Smith");
        stu.setStuNum("12345");
        System.out.println("Student Name: " + stu.getName());
        System.out.println("Student Number: " + stu.getStuNum());
    }
}
```

Here is another example of inheritance.

Java supports only Single Inheritance while C++ also supports Multiple Inheritance.

The Forms of Single Inheritance:

- Single Level Form
- Multiple Level Form
- Hierarchical Form

5.2. SINGLE LEVEL INHERITANCE

In the single level inheritance single class can derived from only one base class. If a class B is derived form class A, like a figure 5.2, then it is called single level inheritance.

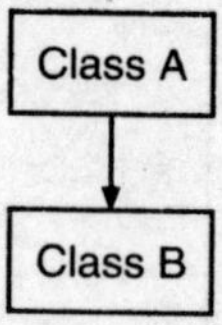

Fig 5.2. Single Level Inheritance

Here we have given some example to understand the concept of single inheritance.

Example 5.1: Single inheritance

```
class single
{
static class A
{
    int a1, a2;
    void getdata()
    {
        a1 = 10;
        a2 = 20;
    }
    void putdata()
    {
        System.out.println("\nThe value of a1 is :"+ a1 + " and a2
:" + a2);
    }
}

static class B extends A            // class B is derived by class A
                          (single level inheritance)
{
    int b1, b2;
    void indata()
    {
        b1 = 30;
        b2 = 40;
    }
    void outdata()
    {
        System.out.println("\nThe value of b1 is :"+ b1 + " and b2
:" + b2);
    }
}

public static void main(String [] args)
```

```
{
B Obj=new B();
Obj.getdata();          // calling base class member function
Obj.indata();           // derived class member function
Obj.putdata();
Obj.outdata();
}
}
```

OUTPUT

```
The value of a1 is : 10  and a2 is : 20

The value of b1 is : 30  and b2 is : 40
```

Example 5.2: Single inheritance with Input Method

```
import java.util.*;
class single
{
static Scanner sc=new Scanner(System.in);
static class A
{
      int a1, a2;
      void getdata()
      {
            System.out.println("Enter the value of a1 :");
            a1 = sc.nextInt();
            System.out.println("Enter the value of a2 :");
            a2 = sc.nextInt();
      }
      void putdata()
      {
            System.out.println("\nThe value of a1 is :"+ a1 + " and a2
:" + a2);
      }
}

static class B extends A              // class B is derived by class A
                                 (single level inheritance)
{
      int b1, b2;
      void indata()
      {
            System.out.println("Enter the value of b1  :");
            b1 = sc.nextInt();
            System.out.println("Enter the value of b2  :");
            b2 = sc.nextInt()
      }
      void outdata()
```

```
    {
            System.out.println("\nThe value of b1 is :"+ b1 + " and b2
:" + b2);
    }
}
public static void main(String [] args)
{
B Obj=new B();
Obj.getdata();  // calling base class member function
Obj.indata();   // derived class member function
Obj.putdata();
Obj.outdata();
}
}
```

OUTPUT

```
Enter the value of a1 :
10
Enter the value of a2 :
20
Enter the value of b1 :
30
Enter the value of b2 :
40

The value of a1 is : 10  and a2 is : 20

The value of b1 is : 30  and b2 is : 40
```

Example 5.3: Single inheritance with protected data

```
import java.util.*;
class single
{
static Scanner sc=new Scanner(System.in);
static class A
{
    protected int a;
    void getdata()
    {
        System.out.println("Enter the value of a ( protected) :");
        a = sc.nextInt();
    }
    void putdata()
    {
        System.out.println("\nThe value of a :" + a);
    }
}
```

```
static class B extends A            // class B is derived by class A
                          (single level inheritance)
{
    int b;
    void indata()
    {
        System.out.println("Enter the value of b:");
        b = sc.nextInt();
    }
    void outdata()
    {
        System.out.println("\nThe value of b is :"+ b);
    }
    void addition()
    {
        int c;
        c = a + b; // a is protected data can be used directly
        System.out.println("\nThe Addition of a & b is :"+ c);
    }
}

public static void main(String [] args)
{
B Obj=new B();

Obj.getdata();          // calling base class member function

Obj.indata();           // derived class member function

Obj.putdata();
Obj.outdata();
Obj.addition();

}
}
```

OUTPUT

```
Enter the value of a :
2

Enter the value of b :
4

The value of a is : 2

The value of b is : 4

Addition of both variables : 6
```

Example 5.4: Single inheritance

```
import java.util.*;
//SINGLE INHERITANCE WITH STUDENT MARKS
class single
{
static Scanner sc=new Scanner(System.in);
static class Student
{
    String name;
    int rollno;
    void getdata()
    {
        System.out.println("\nEnter the name :");
        name=sc.nextLine();
        System.out.println("\nEnter the Roll No. :");
        rollno=sc.nextInt();
    }
    void putdata()
    {
        System.out.println("\n Name  :"+ name);
        System.out.println("\n Roll No. :"+rollno);
    }
}

static class Marks extends Student
{
    int[] marks=new int[4];

void getmarks()
{
for(int i = 0; i < 4 ; i++)
{
System.out.println("\nEnter the Marks of Subject "+(i+1)+" := ");
marks[i]=sc.nextInt();
}
}
void putmarks()
{
int total = 0;
for(int i = 0; i < 4 ; i++)
{
System.out.println("\nMarks of Subject "+ (i+1)+ ": =" +marks[i]);
total = total + marks[i];
}
System.out.println("\n Total Marks :"+total);
}
}
```

```
    public static void main(String [] args)
    {
        Marks M=new Marks();
        M.getdata();
        M.getmarks();
        M.putdata();
        M.putmarks();
    }
}
```

OUTPUT

```
Enter the Name :
Ravi
Enter the Roll No. :
1104
Enter the Marks of Subject 1
65
Enter the Marks of Subject 2
69
Enter the Marks of Subject 3
76
Enter the Marks of Subject 4
54

Name : Ravi
Roll No.: 1104
Marks of Subject 1: = 65
Marks of Subject 2: = 69
Marks of Subject 3: = 76
Marks of Subject 4: = 54
Total Marks : 264
```

5.3. MULTI LEVEL INHERITANCE

In the Multilevel inheritance a class derived from another derived class. If a class B is derived from class A and class C derived from class B, like a figure 5.3 , then it is called multilevel inheritance, because there are more than one levels. You can also derive class C , and then derive its derived and so on.

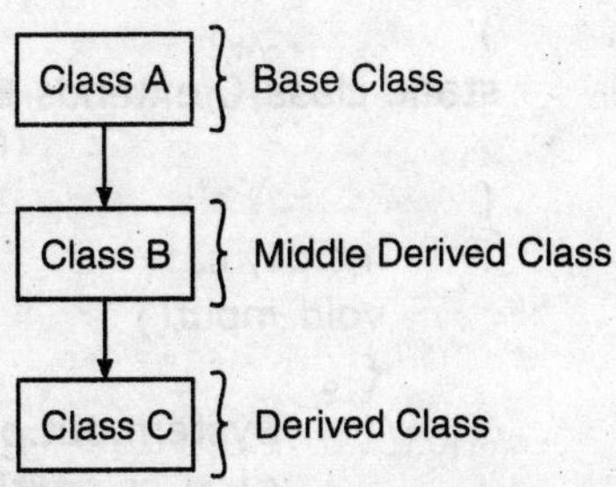

Fig. 5.3. Multilevel Inheritance

Here we have given some example to understand the concept of Multilevel inheritance.

Example 5.5: Progra to Demonstrate Multilevel inheritance

```
import java.util.*;
class multilevel
{
static Scanner sc=new Scanner(System.in);
```

```
static class A
{
    int a1, a2;
    void getdata()
    {
        System.out.println("Enter the value of a1 is :");
    a1 = sc.nextInt();
        System.out.println("Enter the value of a2 is :");
    a2 = sc.nextInt();
    }
    void putdata()
    {
        System.out.println("\nThe value of a1 is :"+ a1 + " and a2
:" + a2);
    }
}

static class B extends A              // class B is derived by class A
{
    int b1, b2;
    void indata()
    {
        System.out.println("Enter the value of b1 is :");
        b1 = sc.nextInt();
        System.out.println("Enter the value of b2 is :");
        b2 = sc.nextInt();
    }
    void outdata()
    {
        System.out.println("\nThe value of b1 is :"+ b1 + " and b2
:" + b2);
    }
}
static class C extends B              // class C is derived by class B
                           (Multi-level inheritance)
{
    int c1, c2;
    void input()
    {
        System.out.println("Enter the value of c1 is :");
        c1 = sc.nextInt();
        System.out.println("Enter the value of c2 is :");
        c2 = sc.nextInt();
    }
    void output()
    {
        System.out.println("\nThe value of c1 is :"+ c1 + " and c2
:" + c2);
    }
```

```
}

public static void main(String [] args)
{
C Obj=new C();
Obj.getdata();  // calling base class member function
Obj.indata();   // Middile derived class member function
Obj.input();           // Derived class member function
Obj.putdata();
Obj.outdata();
Obj.output();
}
}
```

OUTPUT

```
Enter the value of a1  :
2
Enter the value of a2 :
3
Enter the value of b1 :
4
Enter the value of b2 :
5
Enter the value of c1 :
6
Enter the value of c2 :
7

The value of a1 is : 2  and a2  : 3

The value of b1 is : 4  and b2  : 5

The value of c1 is : 6  and c2  : 7
```

Example 5.6: Program of Multilevel inheritance with protected data.

```
import java.util.*;
class multilevel
{
static Scanner sc=new Scanner(System.in);
static class A
{
      protected int a;
      void getdata()
      {
            System.out.println("Enter the value of a ( protected) :");
            a = sc.nextInt();
      }
```

```
        void putdata()
        {
            System.out.println("\nThe value of a :" + a);
        }
    }
    static class B extends A      // class B is derived by class A
    {
        protected int b;
        void indata()
        {
            System.out.println("Enter the value of b (protected):");
            b = sc.nextInt();
        }
        void outdata()
        {
            System.out.println("\nThe value of b is :"+ b);
        }
    }
    static class C extends B      // class C is derived by class B (Multi-
                                  level inheritance)
    {
        int c;
        void input()
        {
            System.out.println("Enter the value of c:");
            c = sc.nextInt();
        }
        void output()
        {
            System.out.println("\nThe value of c is :"+ c);
        }

        void addition()
        {
            int d;
        d = a + b + c;  // a & b is protected data can be used directly
            System.out.println("\nThe Addition of a , b and c is :"+ d);
        }
    }

    public static void main(String [] args)
    {
        C Obj=new C();
        Obj.getdata();  // calling base class member function
        Obj.indata();   // derived class member function
        Obj.input();          // derived class member function
        Obj.putdata();
        Obj.outdata();
```

```
        Obj.output();
        Obj.addition();

    }
    }
```

OUTPUT

```
Enter the value of a :
2
Enter the value of b :
4
Enter the value of c :
6
The value of a is : 2
The value of b is : 4
The value of b is : 6
Addition of both variables : 12
```

Example 5.7: Program of Multilevel inheritance for Student marks Records.

```
import java.util.*;
class multilevel
{
static Scanner sc=new Scanner(System.in);
static class Student
{
        String name;
        int rollno;
        void getdata()
        {
            System.out.println("\nEnter the name :");
            name=sc.nextLine();
            System.out.println("\nEnter the Roll No. :");
            rollno=sc.nextInt();
        }
        void putdata()
        {
            System.out.println("\n Name   :"+ name);
            System.out.println("\n Roll No. :"+rollno);
        }
}

static class Marks extends Student
{
protected int[] marks=new int[4];
void getmarks()
{
for(int i = 0; i < 4 ; i++)
```

```
{
System.out.println("\nEnter the Marks of Subject "+(i+1)+" := ");
marks[i]=sc.nextInt();
}
}
void putmarks()
{
for(int i = 0; i < 4 ; i++)
      System.out.println("\nMarks of Subject "+ (i+1)+ ": =" +marks[i]);
}
}
static class Result extends Marks
{
int total;
float perc;
void Totalmarks()
{
total = 0;
for(int i = 0; i < 4 ; i++)
{
total = total + marks[i];
perc= (float)total*100/400;
}
}
void display()
{
      Totalmarks();
      System.out.println("Total Marks= "+total+"\nPercentage = "+perc);
      if(marks[0]>30 && marks[1]>30 && marks[2]>30 && marks[3]>30)
      {
      if(perc>60F)
            System.out.println("Division: First");
      if(perc<60F && perc>50F)
            System.out.println("Division: Second");
      if(perc<50F && perc>30F)
            System.out.println("Division: Third");
      if(perc<30F)
            System.out.println("Division: Fail");
      }
      else
            System.out.println("Division: Fail");
      }

      }
public static void main(String [] args)
{
            Result R=new Result();
            R.getdata();
```

```
            R.getmarks();
            R.putdata();
            R.putmarks();
            R.display();
    }
    }
```

OUTPUT

```
Enter the Name :
Ravi
Enter the Roll No. :
1104
Enter the Marks of Subject 1
65
Enter the Marks of Subject 2
69
Enter the Marks of Subject 3
76
Enter the Marks of Subject 4
54

Name : Ravi

Roll No.: 1104

Marks of Subject 1: 65
Marks of Subject 2: 69
Marks of Subject 3: 76
Marks of Subject 4: 54
Total Marks :  264
Percentage : 66 %
Division : First
```

5.4. HIERARCHICAL INHERITANCE

In the Hierarchical inheritance more than one class derived from only single base class. If a class B and class C are derived form class A , like a figure 5.4, then it is called hierarchical inheritance. Because there are more than one derived classes of a single base class thus both classes inherits the features of class A.

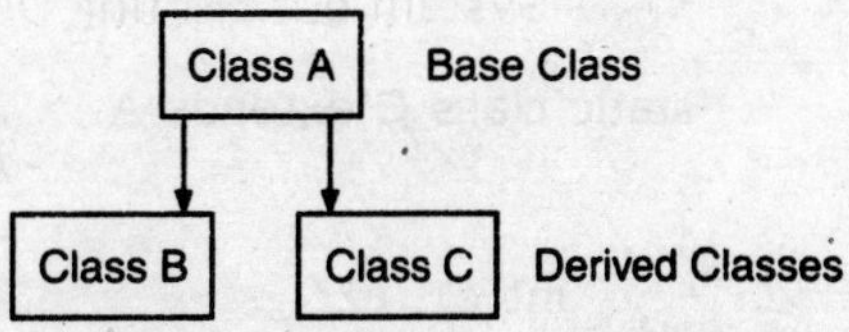

Fig. 5.4. Hierarchical Inheritance

Here we have given some example to understand the concept of Hierarchical inheritance.

Example 5.8: Program to impliment Hierarchical inheritance

```
import java.util.*;
class hierarchical
{
```

```
static Scanner sc=new Scanner(System.in);
static class A
{
      int a1, a2;
      void getdata()
      {
            System.out.println("Enter the value of a1 is :");
            a1 = sc.nextInt();
            System.out.println("Enter the value of a2 is :");
            a2 = sc.nextInt();
      }
      void putdata()
      {
            System.out.println("\nThe value of a1 is :"+ a1 + " and a2
:" + a2);
      }
}

static class B extends A            // class B is derived by class A
{
      int b1, b2;
      void indata()
      {
            System.out.println("Enter the value of b1 is :");
            b1 = sc.nextInt();
            System.out.println("Enter the value of b2 is :");
            b2 = sc.nextInt();
      }
      void outdata()
      {
            System.out.println("\nThe value of b1 is :"+ b1 + " and b2
:" + b2);
      }
}

static class C extends A            // class C is also derived by class
                                   A (Hierarchical inheritance)
{
      int c1, c2;
      void input()
      {
            System.out.println("Enter the value of c1 is :");
            c1 = sc.nextInt();
            System.out.println("Enter the value of c2 is :");
            c2 = sc.nextInt();
      }
      void output()
      {
```

```
        System.out.println("\nThe value of c1 is :"+ c1 + " and c2 :" + c2);
    }
}

public static void main(String [] args)
{
B Ob1 =new B();
C Ob2 =new C();

Ob1.getdata();          // calling by Object of class B
Ob1.indata();

Ob2.getdata();          // calling by Object of class C
Ob2.input();

Ob1.putdata();
Ob1.outdata();

Ob2.putdata();
Ob2.output();
}
}
```

OUTPUT

```
Enter the value of a1  :
1
Enter the value of  a2 :
1
Enter the value of b1 :
2
Enter the value of b2 :
2
Enter the value of a1 :
5
Enter the value of a2 :
5
Enter the value of c1 :
10
Enter the value of c2 :
10

The value of a1 is : 1  and a2 is : 1

The value of b1 is : 2  and b2 is : 2

The value of a1 is : 5  and a2 is : 5

The value of c1 is : 10  and c2 is : 10
```

Example 5.9: Program of Hierarchical inheritance with protected data

```
Import java.util.*;
class hierarchical
{
static Scanner sc=new Scanner(System.in);
static class A
{
    protected int a;
    void getdata()
    {
        System.out.println("Enter the value of a ( protected) :");
        a = sc.nextInt();
    }
    void putdata()
    {
        System.out.println("\nThe value of a :" + a);
    }
}

static class B extends A                // class B is derived by class A
{
    protected int b;
    void indata()
    {
        System.out.println("Enter the value of b :");
        b = sc.nextInt();
    }
    void outdata()
    {
        System.out.println("\nThe value of b is :"+ b);
    }
    void addition()
    {
        int d;
    d = a + b;      // a is protected data can be used directly
        System.out.println("\nThe Addition of a and b is :"+ d);
    }
}
static class C extends A                // class C is also derived by class
                                          A (Hierarchical inheritance)
{
    int c;
    void input()
    {
        System.out.println("Enter the value of c:");
        c = sc.nextInt();
    }
```

```
        void output()
        {
                System.out.println("\nThe value of c is :"+ c);
        }

    void addition()
        {
                int d;
        d = a + c;  // a is protected data can be used directly
                System.out.println("\nThe Addition of a and c is :"+ d);
        }
    }

    public static void main(String [] args)
    {

    B Ob1 =new B();
    C Ob2 =new C();

    Ob1.getdata(); // calling by Object of class B
    Ob1.indata();

    Ob2.getdata(); // calling by Object of class C
    Ob2.input();

    Ob1.putdata();
    Ob1.outdata();
    Ob1.addition();

    Ob2.putdata();
    Ob2.output();
    Ob2.addition();
    }
    }
```

OUTPUT

```
Enter the value of a (protected):
2
Enter the value of b :
4
Enter the value of a (protected):
10
Enter the value of c :
6
The value of a is : 2
The value of b is : 4
Addition of  a and b : 6
```

```
The value of a is : 10
The value of c is : 6
Addition of a and c : 16
```

Example 5.10: Program of Hierarchical inheritance for student marks record.

```
import java.util.*;
class hierarchical
{
static Scanner sc=new Scanner(System.in);
static class Student
{
        String name;
        int rollno;
        void getdata()
        {
            System.out.println("\nEnter the name :");
            name=sc.nextLine();
            System.out.println("\nEnter the Roll No. :");
            rollno=sc.nextInt();
        }
        void putdata()
        {
            System.out.println("\n Name  :"+ name);
            System.out.println("\n Roll No. :"+rollno);
        }
}

static class Marks extends Student
{
protected int[] marks=new int[4];

void getmarks()
{
for(int i = 0; i < 4 ; i++)
{
System.out.println("\nEnter the Marks of Subject "+(i+1)+" := ");
marks[i]=sc.nextInt();
}
}

void putmarks()
{
int total=0;
for(int i = 0; i < 4 ; i++)
{System.out.println("Marks of Subject "+ (i+1)+ ": =" +marks[i]);
total = total + marks[i];
}
```

```
System.out.println("Total Marks= "+total);
}
}

static class Sports extends Student
{
int [] score=new int[3];
void getscore()
{
for(int i = 0; i < 3 ; i++)
{
System.out.println("\nEnter the Score "+(i+1)+" := ");
score[i]=sc.nextInt();
}
}

void totalscore()
{
int total=0;
for(int i = 0; i < 3 ; i++)
{
total = total + score[i];
}
System.out.println("Total Score= "+total);
}
}
public static void main(String [] args)
{
      Marks M=new Marks();
      M.getdata();
      M.getmarks();
      M.putdata();
      M.putmarks();

      Sports S = new Sports();
      S.getdata();
      S.getscore();
      S.putdata();
      S.totalscore();
}
}
```

OUTPUT

```
Enter the Name :
Ravi
Enter the Roll No. :
1104
```

```
Enter the Marks of Subject 1
65
Enter the Marks of Subject 2
69
Enter the Marks of Subject 3
76
Enter the Marks of Subject 4
54
Enter the Name :
Hari
Enter the Roll No. :
1136
Enter the score 1 : 200
Enter the score 2 : 200
Enter the score 3 : 100

Name : Ravi
Roll No.: 1104
Marks of Subject 1:65
Marks of Subject 2:69
Marks of Subject 3:76
Marks of Subject 4:54
Total Marks :  264

Name : Hari
Roll No. : 1136
Total Score : 500
```

5.5. METHOD OVERRIDING

When you call a method on an object, Java looks for that method definition in the class of that object, and if it doesn't find one, it passes the method call up the class hierarchy until a method definition is found. Method inheritance enables you to define and use methods repeatedly in subclasses without having to duplicate the code itself.

> **Concept**
>
> *Method overriding means same name methods in different classes of Inheritance hierchy.*

However, there may be times when you want an object to respond to the same methods but have different behavior when that method is called. In this case, you can override that method. Overriding a method involves defining a method in a subclass that has the same signature as a method in a superclass. Then, when that method is called, the method in the subclass is found and executed instead of the one in the superclass.

5.5.1. Creating Methods that Override Existing Methods

To override a method, all you have to do is create a method in your subclass that has the same signature (name, return type, and parameter list) as a method defined by one of your class's superclasses. Because Java executes the first method definition it finds that matches the signature, this effectively "hides" the original method definition. Here's a simple example 5.11 shows a simple class with a method called display(), which prints out the name of the class from where method is involved.

Example 5.11: Program to demonstrate method overriding.

```
//Declaration of super class
class One
{
public void display()
{
    System.out.println("I am from class one");
}
}
//Declaration of Main class
class override extends One
{
public void display()
{
System.out.println("I am from class  override");
}
public static void main(String [] args)
{
    override o = new override();
    o.display();
}
}
```

OUTPUT

```
I am from class override
```

In the main() method of override class we invoked the display() method, which hides the display() method of class **One** and gives the output from display method of **override** class. Note that to implement the concept the overriding you must use the same name and signature of the method in the inheritance hierarchy. Now see one more example.

Example 5.12: Program to demonstrate method overriding.

```
//Declaration of super class
class One
{
public void display()
{
    System.out.println("I am from class one");
}
}
//Declaration of sub class
class Two extends One
{
```

```
public void display()
{
     System.out.println("I am from class two");
}
}
//Declaration of Main class
class override
{
public void display()
{
     System.out.println("I am from class override");
}
public static void main(String [] args)
{
     override o = new override();
     o.display();
     Two t=new Two();
     t.display();
}
}
```

OUTPUT

```
I am from class override
I am from class two
```

In the above example we have shows that we can also create the object of any class and invoke the method which hidden by its child class. Now question is that can we call the method of base class and super class by the object of leaf class (means last most child class of inheritance hierarchy)? Lets move to next section with more example.

5.5.2. Calling the Original Method

Usually, there are two reasons why you want to override a method that a superclass has already implemented:

- To replace the definition of that original method completely
- To augment the original method with additional behavior

You've already learned about the first one; by overriding a method and giving that method a new definition, you've hidden the original method definition. But sometimes you may just want to add behavior to the original definition rather than erase it altogether. This is particularly useful where you end up duplicating behavior in both the original method and the method that overrides it; by being able to call the original method in the body of the overridden method, you can add only what you need.

To call the original method from inside a method definition, use the super keyword to pass the method call up the hierarchy:

```
void myMethod (String a, String b)
{
     // do stuff here
     super.myMethod(a, b);
     // maybe do more stuff here
}
```

The **super** keyword, somewhat like the this keyword, is a placeholder for this class's superclass. You can use it anywhere you can use this, but to refer to the superclass rather than to the current class.

For example, see those display() methods used in the previous example.

Example 5.13: Program to demonstrate to call original method in overriding concept.

```
//Declaration of base class
class One
{
public void display()
{
    System.out.println("I am from class one");
}
}
//Declaration of sub class
class Two extends One
{
public void display()
{
    super.display();
    System.out.println("I am from class two");
}
}
//Declaration of Main class
class override extends Two
{
public void display()
{
    super.display();
    System.out.println("I am from class override");
}
public static void main(String [] args)
{
    override o = new override();
    o.display();
}
}
```

OUTPUT

```
I am from class one
I am from class two
I am from class override
```

Rather than duplicating most of the behavior of the superclass's method in the sub class, you can rearrange the superclass's method so that additional behavior can easily be added to sub class, as you saw in above example.

5.5.3. Overriding Constructors

Constructors cannot technically be overridden. Because they always have the same name as the current class, you're always creating new constructors instead of inheriting the ones you've got. Much of the time, this is fine, because when your class's constructor is called, the constructor with the same signature for all your superclass is also called, so initialization of all the parts of a class you inherit can happen.

However, when you're defining constructors for your own class, you may want to change how your object is initialized, not only by initializing new variables your class adds, but also to change the contents of variables that are already there. You do this by explicitly calling your superclass's constructors, and then changing whatever you like.

To call a regular method in a superclass, you use super.methodname(arguments). Because with constructors you don't have a method name to call, however, you have to use a different form:

```
super(arg1, arg2, ...);
```

Similar to using this(...) in a constructor, super(...) calls the constructor method for the immediate superclass (which may, in turn, call the constructor of its superclass, and so on).

Example **5**.6, shows a class called NamedPoint, which extends the class Point from Java's awt package. The Point class has only one constructor, which takes an x and a y argument and returns a Point object. NamedPoint has an additional instance variable (a string for the name) and defines a constructor to initialize x, y, and the name.

Example 5.14: The NamedPoint class.

```
import java.awt.Point;
class NamedPoint extends Point
{
        String name;
        NamedPoint(int x, int y, String name)
        {
        super(x,y);
        this.name = name;
        }
}
```

The constructor defined here for NamedPoint (lines 6 through 8) calls Point's constructor method to initialize Point's instance variables (x and y). Although you can just as easily initialize x and y yourself, you may not know what other things Point is doing to initialize itself, so it's always a good idea to pass constructors up the hierarchy to make sure everything is set up correctly.

5.6. ABSTRACT CLASSES AND METHODS

5.6.1. Abstract Classes

Sometimes, a class that you define represents an abstract concept and as such, should not be instantiated. Take, for example, food in the real world. Have you ever seen an instance of food? No. What you see instead are instances of carrot, apple, and (my favorite), chocolate. Food represents the abstract concept of things that we can eat. It doesn't make sense for an instance of food to exist.

Similarly in object-oriented programming, you may want to model abstract concepts but you don't want to be able to create an instance of it. For example, the Number class in the java.lang package represents the abstract concept of numbers. It makes sense to model numbers in a program, but it doesn't make sense to create a generic number object. Instead, the Number class makes sense only as a superclass to classes like Integer and Float which implement specific kinds of numbers. Classes such as Number, which implement abstract concepts and should not be instantiated, are called *abstract classes*. An abstract class is a class that can only be subclassed—it cannot be instantiated.

> **Concept**
> *The abstract class is a class, which can not directly instantiated, its sub-class can be instantiated.*

To declare that your class is an abstract class, use the keyword abstract before the class keyword in your class declaration:

```
abstract class Number
{
    . . .
}
```

If you attempt to instantiate an abstract class, the compiler will display an error similar to the following and refuse to compile your program:

```
AbstractTest.java:6: class AbstractTest is an abstract class.
It can't be instantiated.
new AbstractTest();
^
1 error
```

5.6.2. Abstract Methods

An abstract class may contain *abstract methods*, that is, methods with no implementation. In this way, an abstract class can define a complete programming interface thereby providing its subclasses with the method declarations for all of the methods necessary to implement that programming interface. However, the abstract class can leave some or all of the implementation details of those methods up to its subclasses.

> **Concept**
> *Abstract method is a method without definition. It is implemented by subclasses.*

Let's look at an example of when you might want to create an abstract class with an abstract method in it. In an object-oriented drawing application, you can draw circles, rectangles, lines, Beziers, and so on. Each of these graphic objects share certain states (position, bounding box) and behavior (move, resize, draw). You can take advantage of these similarities and declare them all to inherit from the same parent object—GraphicObject.

However, the graphic objects are also substantially different in many ways: drawing a circle is quite different from drawing a rectangle. The graphics objects cannot share these types of states or behavior. On the other hand, all GraphicObject's must know how to draw themselves; they just differ in how they are drawn. This is a perfect situation for an abstract superclass.

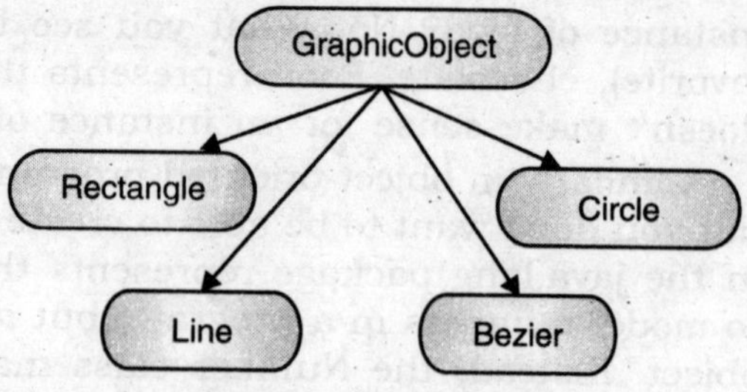

Fig. 5.5. Graphic object

First you would declare an abstract class, GraphicObject, to provide member variables and methods that were wholly shared by all subclasses such as the current position and the moveTo() method. GraphicObject would also declare abstract methods for methods, such as draw(), that needed to be implemented by all subclasses, but implemented in entirely different ways (no suitable default implementation in the superclass makes sense). The GraphicObject class would look something like this:

```
abstract class GraphicObject
{
     int x, y;
     . . .
void moveTo(int newX, int newY)
{
     . . .
}
abstract void draw();
}
```

Each non-abstract subclass of GraphicObject, such as Circle and Rectangle, would have to provide an implementation for the draw() method.

```
class Circle extends GraphicObject
{
     void draw()
     {
          . . .
     }
}
class Rectangle extends GraphicObject
{
     void draw()
     {
          . . .
     }
}
```

An abstract class is not required to have an abstract method in it. But any class that has an abstract method in it or that does not provide an implementation for any abstract methods declared in its superclasses *must* be declared as an abstract class.

Example 5.15: Abstract class

```
abstract class Cat
{
 String abstractClassName = "Cat";
 //abstract classes can have variables
 String getAbstractClassName()
 //abstract classes can have methods
 {
        return abstractClassName;
 }

 abstract String getClassName();
 //this must be implemented in any class extending Cat
}
 class Himalayan extends Cat
 {
 //you can not extend from two classes at once

 String className = "Himalayan";

 public Himalayan() {}

 String getClassName()
 //must have this method,
 // because Cat declared it as an abstract
        {
        return className;
        }

 public static void main(String[] args)
 {

        //Cat percy = new Cat();
        //can not instantiate an abstract class

        Himalayan cappuccino = new Himalayan();

        System.out.println(cappuccino.getAbstractClassName());
        //output is: Cat
        System.out.println(cappuccino.getClassName());
        //output is: Himalayan
 }
 }
```

OUTPUT

```
Cat
Himalayan
```

Example 5.16: Program to display the use of abstract class.

```
// simple abstract class
abstract class Calc
{
        void sum(int a,int b)
        {
                System.out.println("Sum is = "+(a+b));
        }
        abstract void sub(int a,int b);
}
// class inherit with abstract class
class MyCalc extends Calc
{
        void sub(int a,int b) // method of Calc class
        {
                System.out.println("Sub is = "+(a-b));
        }
        public static void main(String aa[])
        {
                MyCalc o = new MyCalc();
                o.sum(100,20);
                o.sub(100,20);
        }
}
```

In the previous sections we have used the classes called concrete classes which always required to use the abstract classes, because abstract classes can't create there instances. There we distinguish both of by table:

Concrete class	Abstract class
only contain concrete methods	can have abstract and concrete methods.
can create instance	can't create instance, inherit only.
can be inherit.	inherit only

The main advantage of abstract classes are this is used for memory management, memory for methods provided by the abstract class.

> **Note :** All abstract methods of an abstract class must be override in sub classes else sub class also a kind of abstract class.

In the future, abstract classes will not in use, because using classes java support only single inheritance, than one class can extends with one concrete class or one abstract class. In the inheritance with concrete class & in the place of abstract class future will use the concept of interfaces. (interface - it is a pure abstract class.)

5.7. PACKAGES

Many times when we get a chance to work on a small project, one thing we intend to do is to put all java files into one single directory. It is quick, easy and harmless.

However if our small project gets bigger, and the number of files is increasing, putting all these files into the same directory would be a nightmare for us. In java we can avoid this sort of problem by using Packages.

Packages are nothing more than the way we organize files into different directories according to their functionality, usability as well as category they should belong to. An obvious example of packaging is the JDK package from SUN (java.xxx.yyy) as shown below in figure 5.6.

> **Concept**
>
> *The package is a container, that can contain the class files, interfaces and subpackages.*

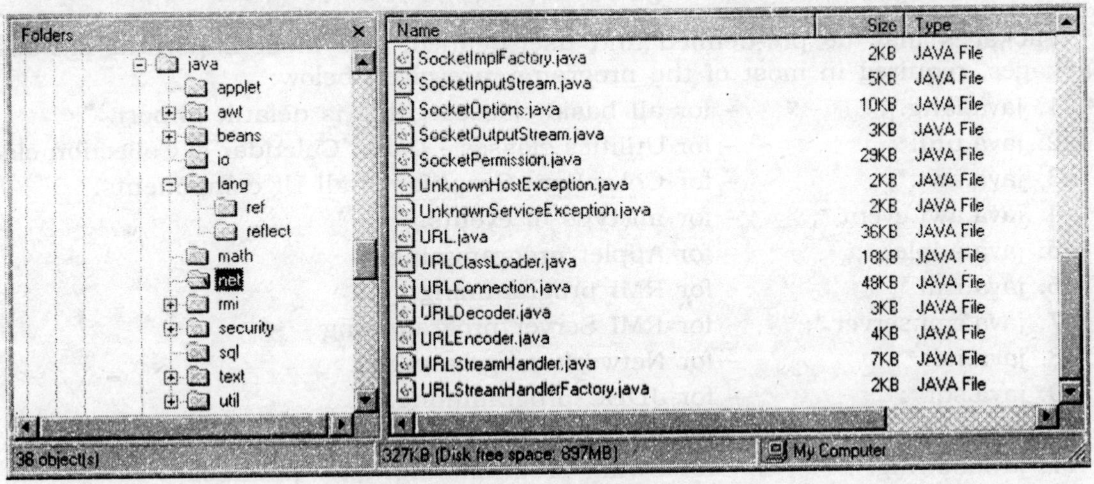

Fig. 5.6. Basic structure of JDK package

There are some comment and definition by some professionals:

- Package is a container, that can contain the class files, interfaces & sub packages.

  – By Harsh Vardhan, M.Tech, IIT Delhi.

- Package is a kind of folder that can contain the class files, interfaces & sub folders.

  – *By Ankit Gupta, Project Leader, Rensole Pvt Ltd, Delhi.*

- Packages are like header files in C & C++, which provides relevance functions to use.

  – *By Deepak Agarwal, Tech Lead, Portech India Info Service Pvt Ltd, Noida.*

- Packages are the bunch of related classes, interface and their functions to provide

  – Simplicity,

  – Reusability,

  – Improved performance,

  – and best for security point of view.

  – *By Dr. R.K. Bhatia, Professor, CSE, Gurukul Kangri University, Haridwar.*

Access modifiers in package

Access property	private	default	protected	public
access in same class	Y	Y	Y	Y
sub class of same package	N	Y	Y	Y
non sub class of same package	N	Y	Y	Y
sub class of another package	N	N	Y	Y
non sub class of another package	N	N	N	Y

Note: protected data members can access in sub class of another package but not access in non sub class of another package.

Packages may be pre-defined and user-defined. The most important pre-defined packages, required in most of the programs, are given below:

1. java.lang.*; – for all basic classes , this is default import.
2. java.util.*; – for Utilities classes - Date , Calendar & Collection classes
3. java.awt.*; – for Color,Font,Graphics & all UI components.
4. java.awt.event.*; – for all type of events.
5. java.applet.*; – for Applet programming.
6. java.rmi.*; – for RMI programming.
7. java.rmi.server.*; – for RMI Server programming.
8. java.net.*; – for Network programming.
9. java.sql.*; – for JDBC programming.
10. java.io.*; – for I/O classes
11. javax.swing.*; – for Swing programming. ,(x - extended)
12. javax.swing.event.*; – for Swing Events.
13. javax.swing.table.*; – for Swing table.
14. javax.servlet.*; – for Servlet programming.
15. javax.servlet.http.*; – for http Servlet programming.

5.7.1. How to create a package

Suppose we have a file called HelloWorld.java, and we want to put this file in a package **world**. First thing we have to do is to specify the keyword **package** with the name of the package we want to use (**world** in our case) on top of our source file, before the code that defines the real classes in the package, as shown in our HelloWorld class below:

```
package world;

public class HelloWorld {
        public static void main(String[] args) {
        System.out.println("Hello World");
        }
}
```

One thing you must do after creating a package for the class is to create nested subdirectories to represent package hierachy of the class. In our case, we have the

world package, which requires only one directory. So, we create a directory **world** and put our HelloWorld.java into it.

Right now we have HelloWorld class inside world package. Next,. we have to introduce the world package into our **CLASSPATH**.

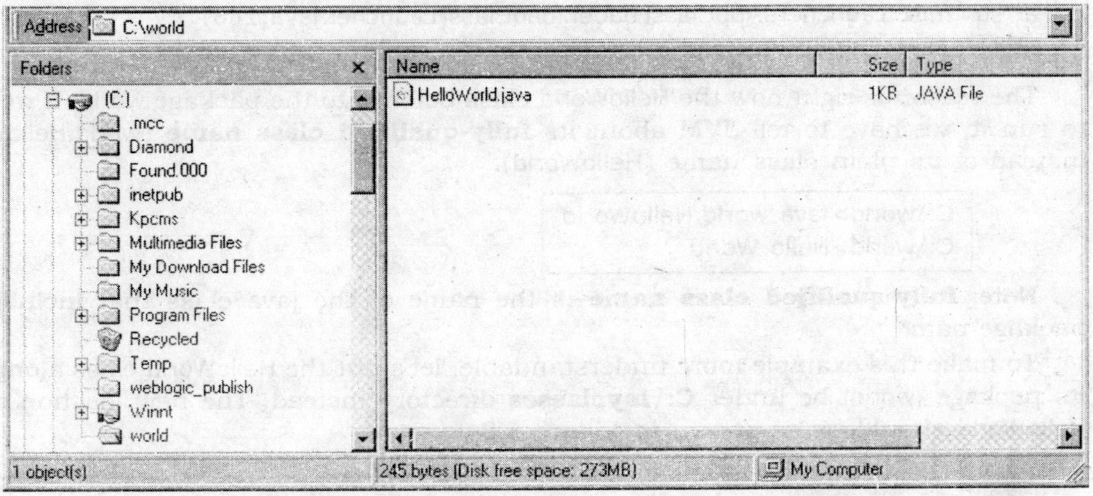

Fig. 5.7. HelloWorld in world package (C:\world\HelloWorld.java)

5.7.2. Setting up the CLASSPATH

From figure **5**.4 we put the package world under C:. So we just set our CLASSPATH as:

```
set CLASSPATH=.;C:\;
```

We set the CLASSPATH to point to 2 places, . (dot) and C:\ directory.

Tip: If you used to play around with DOS or UNIX, you may be familiar with . (dot) and .. (dot dot). We use . as an alias for the current directory and .. for the parent directory. In our CLASSPATH we include this . for convenient reason. Java will find our class file not only from C: directory but from the current directory as well. Also, we use ; (semicolon) to separate the directory location in case we keep class files in many places.

When compiling HelloWorld class, we just go to the world directory and type the command:

```
C:\world\javac HelloWorld.java
```

If you try to run this HelloWorld using **java HelloWorld**, you will get the following error:

```
C:\world>java HelloWorld
Exception in thread "main" java.lang.NoClassDefFoundError: HelloWorld (wrong name:
world/HelloWorld)
at java.lang.ClassLoader.defineClass0(Native Method)
at java.lang.ClassLoader.defineClass(ClassLoader.java:442)
at java.security.SecureClassLoader.defineClass(SecureClassLoader.java:101)
at java.net.URLClassLoader.defineClass(URLClassLoader.java:248)
at java.net.URLClassLoader.access$1(URLClassLoader.java:216)
```

```
at java.net.URLClassLoader$1.run(URLClassLoader.java:197)
at java.security.AccessController.doPrivileged(Native Method)
at java.net.URLClassLoader.findClass(URLClassLoader.java:191)
at java.lang.ClassLoader.loadClass(ClassLoader.java:290)
at sun.misc.Launcher$AppClassLoader.loadClass(Launcher.java:286)
at java.lang.ClassLoader.loadClass(ClassLoader.java:247)
```

The reason is right now the HelloWorld class belongs to the package world. If we want to run it, we have to tell JVM about its **fully-qualified class name** (world.HelloWorld) instead of its plain class name (HelloWorld).

```
C:\world>java world.HelloWorld
C:\world>Hello World
```

Note: **fully-qualified class name** is the name of the java class that includes its package name

To make this example more understandable, let's put the HelloWorld class along with its package (world) be under **C:\myclasses** directory instead. The new location of our HelloWorld should be as shown in Figure **5**.8.

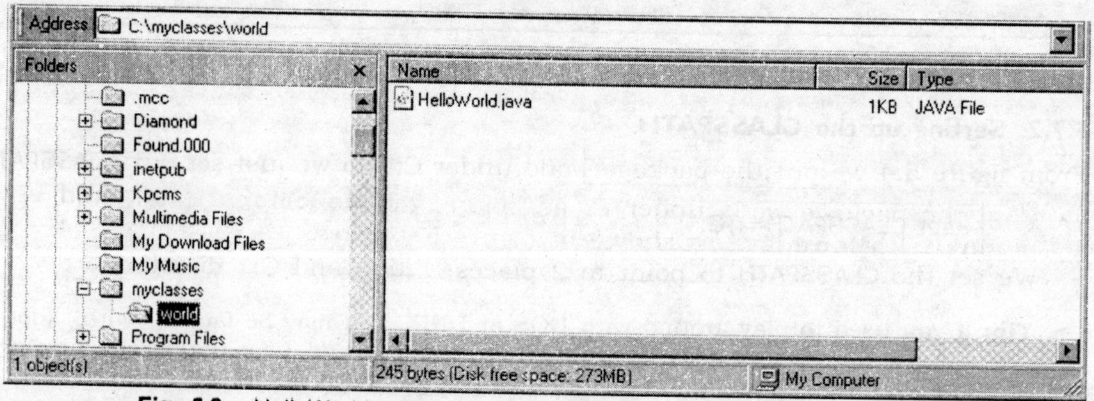

Fig. 5.8. HelloWorld class (in world package) under myclasses directory

We just changed the location of the package from C:\world\HelloWorld.java to C:\myclasses\world\HelloWorld.java. Our CLASSPATH then needs to be changed to point to the new location of the package world accordingly.

```
set CLASSPATH=.;C:\myclasses;
```

Thus, Java will look for java classes from the current directory and **C:\myclasses** directory instead.

Someone may ask "Do we have to run the HelloWorld at the directory that we store its class file everytime?". The answer is NO. We can run the HelloWorld from anywhere as long as we still include the package world in the CLASSPATH. For example,

```
C:\>set CLASSPATH=.;C:\;
C:\>set CLASSPATH // see what we have in CLSSPATH
CLASSPATH=.;C:\;
C:\>cd world
C:\world>java world.HelloWorld
```

```
Hello World
C:\world>cd ..
C:\>java world.HelloWorld
Hello World
```

5.7.3. Subpackage (package inside another package)

Assume we have another file called **HelloMoon.java**. We want to store it in a subpackage **"moon"**, which stays inside package **world**. The HelloMoon class should look something like this:

```
package world.moon;

public class HelloMoon
{
  private String holeName = "rabbit hole";

  public getHoleName() {
  return hole;
  }

  public setHole(String holeName) {
  this.holeName = holeName;
  }
}
```

If we store the package world under C: as before, the **HelloMoon.java** would be **c:\world\moon\HelloMoon.java** as shown in Figure 5.9 below:

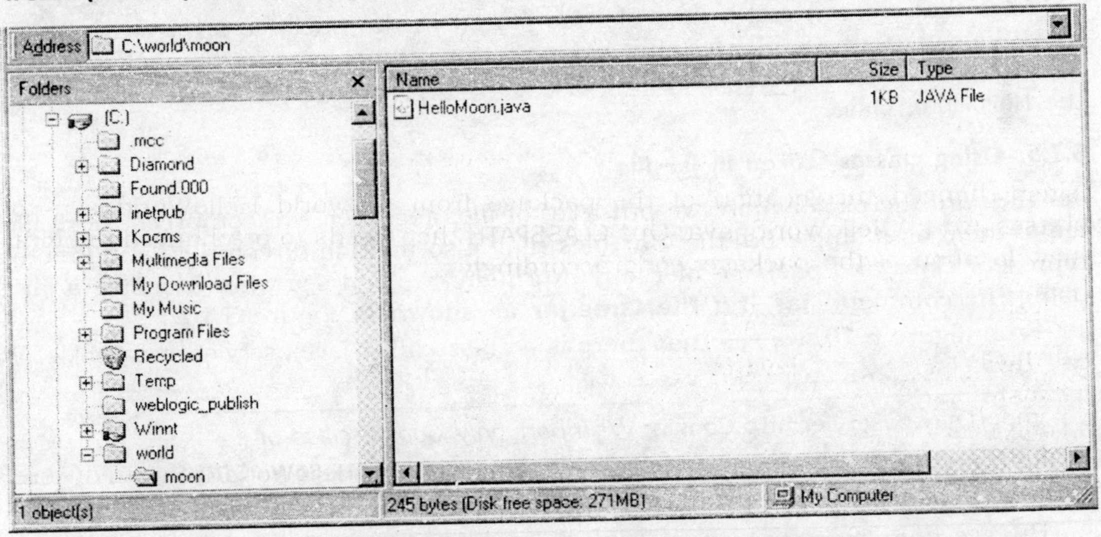

Fig. 5.9. HelloMoon in world.moon package

Although we add a subpackage under package **world**, we still don't have to change anything in our CLASSPATH. However, when we want to reference to the HelloMoon class, we have to use world.moon.HelloMoon as its fully-qualified class name.

5.7.4. How to use package

There are 2 ways in order to use the public classes stored in package.

1. Declare the fully-qualified class name. For example,

```
...
world.HelloWorld helloWorld = new world.HelloWorld();
world.moon.HelloMoon helloMoon = new world.moon.HelloMoon();
String holeName = helloMoon.getHoleName();
...
```

2. Use an "import" keyword:

```
import world.*;              // we can call any public classes inside the world packag
import world.moon.*;         // we can call any public classes inside the world.moon
                             package
import java.util.*;          // import all public classes from java.util package
import java.util.Hashtable;

                             // import only Hashtable class (not all classes in java.util
                             package)
```

Thus, the code that we use to call the HelloWorld and HelloMoon class should

```
...
HelloWorld helloWorld = new HelloWorld();
 // don't have to explicitly specify world.HelloWorld anymore
HelloMoon helloMoon = new HelloMoon();
 // don't have to explicitly specify world.moon.HelloMoon anymore
...
```

Note that we can call public classes stored in the package level we do the impo only. We can't use any classes that belong to the subpackage of the package we impor For example, if we import package world, we can use only the HelloWorld class, but n the HelloMoon class.

5.7.5. Using classes stored in jar file

Jar files are the place where we put a lot of files together. We compress these files an make them as a single bundle. Jar files may also include directories, subdirectories t represent class and package hierachy. Normally, we can see what is inside a jar file b using the command **jar -tvf fileName.jar** as shown in Figure 5.10 :

From figure 5.10. we see that there is a class called javax.servlet.http.Cookie. We ca call this class by

```
import javax.servlet.http.Cookie; // import only Cookie class or
import javax.servlet.http.*;      // import the whole javax.servlet.http
                                     package
```

But we have to include this package in the CLASSPATH as well.

```
set CLASSPATH=.;D:\JSDK2.0\lib\jsdk.jar;
```

Note that if the package is stored inside a jar file, we have to include the jar fil with its extension (.jar) in the CLASSPATH. However, if the package is a plain directory we just put the name of directory into the CLASSPATH.

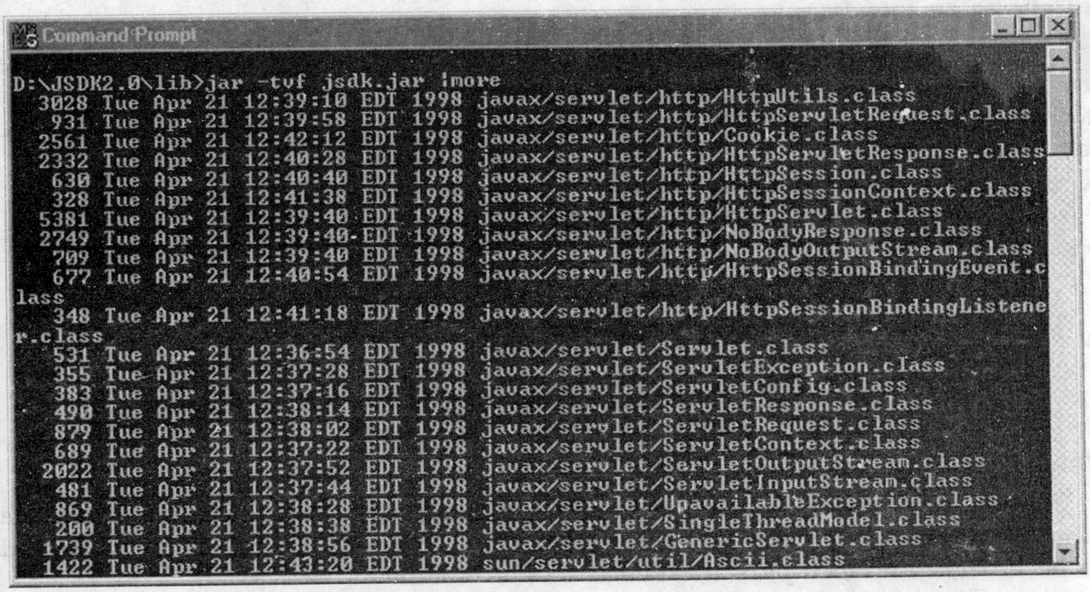

Fig. 5.10. JSDK package viewed by jar -tvf jsdk.jar command

5.8. INTERFACE

English, an *interface* is a device or system that unrelated entities use to interact. cording to this definition, a remote control is an interface between you and a television t, the English language is an interface between two people, and the protocol of havior enforced in the military is the interface between people of different ranks. milarly, a Java interface is a device that unrelated objects use to interact with one other. Java interfaces are probably most analogous to protocols (an agreed-upon havior). In fact, other object-oriented languages have the functionality of Java's terfaces, but they call their interfaces *protocols*.

A Java interface defines a set of methods but does not implement them. A class that plements the interface agrees to implement all of the methods defined in the interface, ereby agreeing to certain behavior.

> **Concept**
>
> *An interface is a named collection of method diclarations (without definitions). An interface can also include constant declarations.*

There are some comment and definition by some professionals:

- Interface is the collection of abstract methods & final variables.

  – By Ankit Gupta, Project Leader, Rensole Pvt Ltd, Delhi.
- Interface is a kind of abstract class that can contain only abstract methods & final variable.

  – By Harsh Vardhan, M.Tech, IIT Delhi.
- Interface is a pure abstract class.

  – By Deepak Agarwal, Tech Lead, Portech India Info Service Pvt Ltd, Noida

5.8.1. Defining an Interface

The following figure shows that an interface definition has two components: the interface declaration and the interface body. The interface declaration declares various attributes about the interface such as its name and whether it extends another interface. The interface body contains the constant and method declarations within that interface.

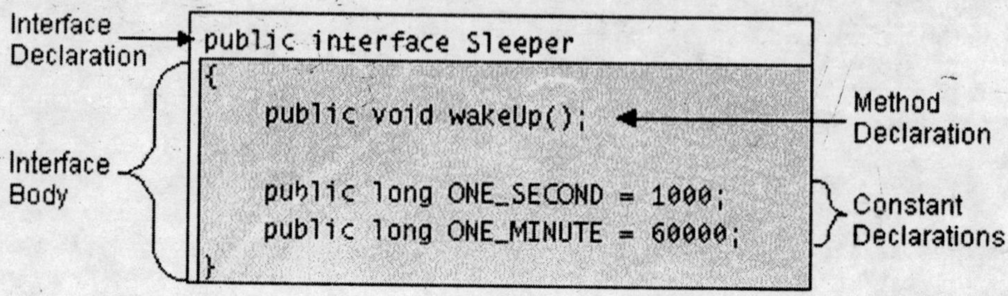

The Interface Declaration

An interface is primarily created like a class, the declaration for the Sleeper interface uses the two required elements of an interface declaration—the interface keyword and the name of the interface—plus the public access modifier :

```
public interface Sleeper
{
. . .
}
```

An interface declaration can have one other component: the public access specifier and a list of superinterfaces. The full interface declaration looks like this figure:

`public`	Makes this interface public.
`interface InterfaceName`	Class cannot be instantiated.
`Extends SuperInterfaces`	This interface's superinterfaces.
`{ InterfaceBody }`	

The public access specifier indicates that the interface can be used by any class in any package. If you do not specify that your interface is public, then your interface will be accessible only to classes that are defined in the same package as the interface.

An interface can extend other interfaces just as a class can extend or subclass another class. However, while a class can extend only one other class, an interface can extend any number of interfaces. The list of superinterfaces is a comma-delimited list of all of the interfaces extended by the new interface. An interface inherits all constants and methods from its superinterfaces, unless the interface hides a constant with another of the same name or redeclares a method with a new method declaration.

The interface body

The interface body contains method declarations for all of the methods included in the interface. Here's the body of the Sleeper interface:

```
public interface Sleeper {
        public void wakeUp();

        public long ONE_SECOND = 1000;
        public long ONE_MINUTE = 60000;
}
```

The method declaration for wakeUp is followed by a semicolon (;) because an interface does not provide implementations for the methods declared within it. All methods declared in an interface are implicitly public and abstract. The use of these modifiers on a method declaration in an interface is discouraged as a matter of style.

An interface can contain constant declarations in addition to method declarations. The Sleeper interface declares two constants. All constant values defined in an interface are implicitly public, static, and final. The use of these modifiers on a constant declaration in an interface is discouraged as a matter of style.

Any class can use an interface's constants from the name of the interface, like this:

```
Sleeper.ONE_SECOND
```

An interface is mostly used to lay a foundation for other classes. For this reason, it is the prime candidate for class derivation. To derive from an interface, you must use the keyword ***implements*** as shown in example. Here is an example of a class named SportBall that derives from an interface named ISportType:

```
public class SportBall implemets ISportType
{
        int players;
        string sport;
}
```

Just as you can derive a class from an interface, you can create an interface that itself is based on another interface. Here is an example:

```
public interface ISportType extends IBall
{

}
```

Java doesn't allow multiple inheritance which is the ability to create a class based on more than one class. Multiple inheritance is allowed only if the bases are interfaces. To create multiple inheritance, separate the names of interface with a comma. Here is an example:

```
public interface ISportType extends IBall, ICourtDimensions
{

}
```

Here is some example program to implement the interface, as well as multiple inheritance:

Example 5.17: Program to use simple interface Calc in the main class.

```
// simple interface
interface Calc
{
     void sum(int a,int b);
     void sub(int a,int b);
}
// class implements with Interface
class MyCalc implements Calc
{
     public void sum(int a,int b) // method of Calc
     {
          System.out.println("Sum is ="+(a+b));
     }
     public void sub(int a,int b) // method of Calc
     {
          System.out.println("Sub is = "+(a-b));
     }
     public static void main(String aa[])
     {
          MyCalc o = new MyCalc();
          o.sum(100,20);
          o.sub(100,20);
     }
}
```

OUTPUT

```
Sum is = 120
Sub is = 80
```

Now you may be little bit confuse about interface and abstract class, so don't worry here are some differences between them:

Abstract class	Interface
It contain abstract and concrete methods.	It contain only abstract methods.
Using abstract classes java support only single inheritance.	Using interfaces java support multiple inheritance also.
All attributes of class by default is normal.	All attributes of interface by default is final.
Default access modifier of abstract class is default.	Default access modifier of interface is public.
It can have constructors.	Interface doesn't have constructors.
It uses extends keyword for inheritance.	It uses implements keyword for inheritance

✎ **Note:** when you will use the concept of inheritance you must remember the which keywords to use where, see the following table:

Inheritance	Keyword
class to class	extends keyword.
interface to interface	extends keyword
interface to class	implements keyword.

There are some restrictions while we are using interface, which are given below:

- All methods of an interface must be override in child class else child class is a kind of abstract class.
- All methods of an interface must be override with public access modifier else we have error - weaker access privileges - was public.

Example 5.18: Simple Interface

```
interface Printem
{
 void printText();
}

class class1 implements Printem
{
 public void printText()
 {
 System.out.printin("Hello from Java!");
 }
}

public class interfaces
{
 public static void main(String[] args)
 {
 class1 object1 = new class1();
 object1.printText();
 }
}
```

Example 5.19: Implementation of multiple interfaces

```
interface Add
{
 public int sum();
}

interface Multiply
{
 public int mul();
}
```

```
class calculate implements Add, Multiply
{
 int a,b;
 public calculate( int x, int y)
 {
      a = x;
      b = y;
 }

 public int sum()
 {
      return ( a + b );
 }
      public int mul()
 {
      return ( a * b );
 }
}

class MyInterface
{
 public static void main(String[] arg)
 {

      calculate cal = new calculate(5,10);

      Add A = (Add) cal;            // casting
      System.out.println("Sum : " + A.sum());

      Multiply M = (Multiply) cal;
      System.out.println("Multiplication :" + M.mul());
 }
}
```

OUTPUT

```
Sum : 15
Multiplication : 50
```

Example 5.20: Multiple implementations of an interface

```
interface calculate
{
 public int compute( int x, int y);
}

class Add implements calculate
{
 public int compute(int a,int b)
```

```
    {
        return ( a + b );
    }
  }

class Multiply implements calculate
{
  public int compute(int a,int b)
  {
        return ( a * b );
  }
}

class NewInterface
{
  public static void main(String[] arg)
  {

        Add A = new Add();
        Multiply M =new Multiply();

        calculate cal;

        cal = (calculate) A; //casting
        System.out.println("Sum : " + cal.compute(2,3));

        cal = (calculate) M; //casting
        System.out.println("Multiplication :" + cal.compute(2,3));
  }
}
```

OUTPUT

```
Sum : 5
Multiplication : 6
```

Example 5.21: Interface

```
interface Fruit
{
 public boolean hasAPeel();
 //has a peel must be implemented in any class implementing Fruit
 //methods in interfaces must be public
}

interface Vegetable
{
 public boolean isARoot();
 //is a root must be implemented in any class implementing Vegetable
 //methods in interfaces must be public
```

```
}
class FandVUtility
{
 static String doesThisHaveAPeel(Fruit fruitIn)
 {
 if (fruitIn.hasAPeel())
 return "This has a peel";
 else
 return "This does not have a peel";
 }
 static String isThisARoot(Vegetable vegetableIn)
 {
 if (vegetableIn.isARoot())
 return "This is a root";
 else
 return "This is not a root";
 }
 static String doesThisHaveAPeelOrIsThisRoot(Tomato tomatoIn)
 {
 if (tomatoIn.hasAPeel() && tomatoIn.isARoot())
 return "This both has a peel and is a root";
 else if (tomatoIn.hasAPeel() || tomatoIn.isARoot())
 return "This either has a peel or is a root";
 else
 return "This neither has a peel or is a root";
 }
}

class Tomato implements Fruit, Vegetable
{
boolean peel = false;
 boolean root = false;

 public Tomato() {}

 public boolean hasAPeel()
 //must have this method,
 // because Fruit declared it
 {
 return peel;
 }
 public boolean isARoot()
 //must have this method,
 // because Vegetable declared it
 {
 return root;
 }
```

```
public static void main(String[] args)
{
//Part one: making a tomato
Tomato tomato = new Tomato();

System.out.println(FandVUtility.isThisARoot(tomato));
//output is: This is not a root

System.out.println(FandVUtility.doesThisHaveAPeel(tomato));
//output is: This does not have a peel
 System.out.println(FandVUtility.doesThisHaveAPeelOrIsThisRoot(tomato));
//output is: This neither has a peel or is a root

//Part two: making a fruit
//Fruit = new Fruit();
//can not instantiate an interface like
// this because Fruit is not a class

Fruit tomatoFruit = new Tomato();
//can instantiate by interface like
// this because Tomato is a class

//System.out.println(FandVUtility.isThisARoot(tomatoFruit));
//can not treat tomatoFruit as a Vegetable
// without casting it to a Vegetable or Tomato
System.out.println(FandVUtility.doesThisHaveAPeel(tomatoFruit));
//output is: This does not have a peel
//
 System.out.println(FandVUtility.doesThisHaveAPeelOrIsThisRoot(tomatoFruit));
//can not treat tomatoFruit as a Vegetable
// without casting it to a Vegetable or Tomato
}
}
```

5.9. THE INSTANCEOF OPERATOR

As discussed before you can pass objects around using references to their parent classes, sometimes you might want to know what actual objects you have. This is the pupose of the instanceof operator. Suppose the class hierarchy is extended so that you have the following:

```
public class Employee extends Object
{
}
public class Manager extends Employee
{
}
public class Engineer extends Employee
{
}
```

If you receive an object using a reference of type Employee, it might turn out to be a Manager or Engineer. You can test it by using instanceof as follows:

```
public void function(Employee e)
{
    if (e instanceof Manager)
    {
        //process for Manager
    }
    else if (e instanceof Engineer)
    {
        //process for Engineer
    }
    else
    {
        //process for any other type of employee
    }
}
```

5.9.1. Casting Objects

In circumstances where you have received a reference to a parent class, and you are determined that the object is in fact, a particular subclass by using the instanceof operator, you can access the full functionality of the objects by casting the reference.

```
public void function(Emplyee e)
{
    if (e instanceof Manager)
{
    Manager m = (Manager) e;
    System.out.println("This is the manager of" + m.getDepartment());
    }
    // rest part of function
}
```

If you do not make the cast, an attempt to execute m.getDepartment() wouldFail because the compiler cannot locate a method called getDepartment in the Employee class.

If you do not make the test using instanceof, you run the risk of the cast failing. Generally, any attempt to cast an object reference is subjected to several checks:

- Casts *upwards* in the class hierarchy are always permitted and, in fact, do not require the cast operator. They can be done by simple assignment.
- For *downward* casts, the compiler must be satisfied that the cast is at least possible. For Example, any attempt to cast a Manager reference to a Engineer reference is not permitted, because the Engineer is not a Manager. The class to which the cast is taking place must be some subclass of the current reference type.
- If the compiler permits the cast, then the object type is checked at runtime. For example, if it turns out that the instanceof check is omitted from the source, and the object being cast is not in fact an object of the type it is being cast to, then a runtime error (exception) occurs.

5.10. THE OBJECT CLASS

The Object class is the root of all classes in the Java technology programming language. If a class is declared with no extends clause, then the compiler adds implicitly the code extends Object to the declaration; for example:

```
 public class Employee
 {
 }
is equivalent to:
 public class Employee extends Object
 {
 }
```

5.10.1. The equals Method

The == operator performs an equivalent comparison. That is , for any reference values x and y, x==y returns true if and only is x and y refer to the same object.

The Object class in the java.lang package has the method public boolean equals(Object obj), which compares two objects for equality. When not overridden, an object's equals() method returns true only if the two references being compared refer to the same object. However, the intention of the equals() method is to compare the contents of two objects whenever possible. This is why the method is frequently overridden. For example, the equals() method in string class returns true if and only if the argument is not null and is a String object that represents the same sequence of characters as the String object with which the method is invoked.

The following example , the MyDate class has modified to indicate an equals method that tests against the year, month, and day attributes.

Example 5.22: An equal example

```
class MyDate
{
 private int day;
 private int month;
 private int year;

public MyDate(int day, int month, int year)
{
  this.day = day;
  this.month=month;
  this.year=year;
}

public boolean equals(Object o)
{
  boolean result = false;
  if ((o!=null) && (o instanceof MyDate))
  {
```

```
        MyDate d =(MyDate)o;
        if ((day ==d.day)&&(month==d.month)&&(year==d.year))
        {
            result=true;
        }
    }
    return result;
   }
   public int hashcode()
   {
    return(day^month^year);
   }
   }
   class TestEqual
   {
   public static void main(String [] arg)
   {
    MyDate date1 = new MyDate(14,3,1976);
    MyDate date2 = new MyDate(14,3,1976);

    if(date1==date2)
          System.out.println("date1 is identical to date2");
    else
          System.out.println("date1 is not identical to date2");

    if(date1.equals(date2))
          System.out.println("date1 is identical to date2");
    else
          System.out.println("date1 is not identical to date2");

    System.out.println("set date2 = date1");

   date2=date1;

    if(date1==date2)
          System.out.println("date1 is identical to date2");
    else
          System.out.println("date1 is not identical to date2");
   }
   }
```

OUTPUT

```
date1 is not identical to date2
date1 is equal to date2
set date2 = date1
date1 is identical to date2
```

5.10.2. The toString Method

The toString method converts an object to a String representation. It is referenced by the compiler when automatic string conversion takes place. For example, the System.out.println() call:

```
Date now =new Date();
System.out.println(now);
```

is equivalent to:

```
System.out.println(now.toString());
```

The Object class defines a default toString method that returns the class name and its reference address (not normally useful). Many classes override toString to provide more useful information. For example, all wrapper classes (introduced in next section) override toString to provide a string form of the value they represent. Even classes representing items without a string form often implement toString to return object state information for debugging purposes.

5.11. WRAPPER CLASSES

The Java programming language does not look at primitive data types as objects. For Example, numerical, Boolean and character data are treated in the primitive form for the sake of efficiency. The Java programming language provides wrapper classes to manipulate primitive data elements as objects. Such data elements are wrapped in an object created around them. Each Java primitive data type has a corresponding wrapper class in the java.lang package. Each wrapper class object encapsulates a single primitive value as shown in given table :

Primitive Data Type	Wrapper Class
boolean	Boolean
byte	Byte
char	Character
short	Short
int	Integer
long	Long
float	Float
double	Double

5.11.1. Boxing

If you have to change the primitive data types to object equivalents (called boxing), then you need to use the wrapper classes. Also to get the primitive data type from the object reference (called unboxing), you need to use the wrapper class methods.

Concept

*The conversion of primitive data type to object type is called **boxing**, and conversion to object type to primitive data type is called **unboxing**.*

You can construct a wrapper class object by passing the value to be wrapped into the appropriate constructor, as shown in following code:

```
int a =420;
Integer b =new Integer(a);  // this is called boxing
int c =b.intValue();        //this is called unboxing
```

Wrapper classes are useful when converting primitive data types because of the many wrapper class methods available; for example:

```
int x = Integer.valueOf(str).intValue();
```

or:

```
int x = Integer.parseInt(str);
```

5.11.2. Autoboxing of Primitive Types

All of this boxing and unboxing can clutter up your code and thus make your code difficult to understand. In J2SE version 5.0, the autoboxing feature enables you to assign and retrieve primitive types without the need of wrapper classes.

Following code shows two simple cases of autoboxing and autounboxing. Compare this code with previous boxing and unboxing.

```
int a = 420;
Integer b = a;     // this is called auto-boxing
int c = b;         //this is called auto-unboxing
```

The J2SE version 5.0 compiler will now create the wrapper object automatically when assigning a primitive to a variable of the wrapper class type. The compiler will also extract the primitive value when assigning from a wrapper object to a primitive variable.

This can be done when passing parameters to methods or even within arithmetic expression.

SUMMARY

Java is a pure object oriented language and all object oriented programming based on the classes and objects and in we can't create any single program without class.

- Java supports Inheritance but not directly support multiple inheritance.
- Java has also the concept of abstract class and abstract methods.
- Abstract classes cannot directly be instantiated, its subclasses can be instantiated.
- Abstract method is a method with no implementation, its implementation done by the subclasses.
- Java has given a concept of package to remove the collision of class name.
- Java has no multiple inheritance but we can implement the concept of multiple inheritance using interface.

SOLVED PROBLEMS

Q.5.1. What's the difference between abstraction and encapsulation?

Ans. Abstraction and encapsulation are two quite separate concepts in Java. Abstraction is a technique that is used to represent common functionality amongst a set of classes. An analogy is to look at a set of vehicles: Car, Truck and Bus have some common features that all road vehicles share. From a Java perspective, the mechanics of turning four wheels, an engine, could be handled in abstract form through a common superclass. Abstraction allows subclasses to share common code without duplication. Encapsulation refers to binding the attributes and behaviours in a single unit.

Q.5.2. What do you mean by Abstract classes, abstract methods?

Ans. Abstract classes are Java classes which are declared as abstract and cannot be instantiated. They can contain non-abstract as well as abstract methods. Abstract methods are methods which are declared with the keyword "abstract" and their definition is deferred to the subclasses. A class having at least one abstract method must be declared abstract.

Q.5.3. What are interfaces? or How they support multiple inheritance in Java?

Ans. Interfaces are classes which are declared with the keyword interface and contain only abstract methods and static variables. The abstract methods of an interface are by default public and so are the variables. Since Java does not support multiple inheritance as C++, interfaces are a way by which we can achieve it in Java. A class can implement multiple interfaces. However, it must provide the implementation of all the abstract methods of the interfaces it has inherited.

Q.5.4. Why Java not directly supports Multiple Inheritance?

Ans. Java does not support Multiple Inheritence because, consider an example: A class named 'Super' has a method display(), we are deriving two classes 'SuperA & SuperB' from the 'Super' class. Now both SuperA & SuperB has the method display().

Now you are deriving a class 'SubOfAB' from 'SuperA' & 'SuperB', then how the class 'SubOfAB' uses the method display()-

(*a*).through the class SuperA (or)

(*b*).through the class SuperB

Due to this drawback Java does not have multiple inheritance.

Q.5.5. If a method is declared as protected, where may the method be accessed?

Ans. A protected method may only be accessed by classes or interfaces of the same package or by subclasses of the class in which it is declared.

Q.5.6. Is println overloading or overriding?

Ans. The println() method is an example of overloading because several methods in the class have the same name and return the same type. In this case, println(boolean), println(int) and println(String) all have the same basic method name, "println", and all return void. The only part of the method signature that varies is the type of the argument (including none), which is enough for the Java interpreter to identify the appropriate method to call at runtime.

Q.5.7. Why can't a class be declared private?

Ans. If a class were declared private, no one could access it, including the runtime. Therefore the class could never be instantiated.

Q.5.8. If you can add methods and fields to a subclass, why can't you remove them?

Ans. If you were to remove methods and fields from a subclass, then the subclass could no longer be used as an instance of its superclass. In fact the subclass might not have any relation at all to the superclass if all the superclass's methods and fields were removed.

Q.5.9. What's the basic difference between an abstract class and an interface?

Ans. An interface can be implemented by a class which is already a subclass of another class. A class which is already a subclass of another class cannot extend an abstract class as well. Furthermore, an abstract class can contain non-abstract methods. An interface cannot.

Q.5.10. What is the package?

Ans. The package is a Java namespace or part of Java libraries. The Java API is grouped into libraries of related classes and interfaces; these libraries are known as packages.

Q.5.11. What is an enumeration?

Ans. An enumeration is an interface containing methods for accessing the underlying data structure from which the enumeration is obtained. It is a construct which collection classes return when you request a collection of all the objects stored in the collection. It allows sequential access to all the elements stored in the collection.

Q.5.12. What is the Vector class?

Ans. The Vector class provides the capability to implement a growable array of objects.

Q.5.13. What is the Set interface?

Ans. The Set interface provides methods for accessing the elements of a finite mathematical set.Sets do not allow duplicate elements.

Q.5.14. What is Dictionary class?

Ans. The Dictionary class is the abstarct super class of Hashtable and Properties class.

Dictionary provides the abstarct functions used to store and retrieve objects by key-value.This class allows any object to be used as a key or value.

Q.5.15. What is the Hashtable class?

Ans. The Hashtable class implements a hash table data structure. A hash table indexes and stores objects in a dictionary using hash codes as the objects' keys. Hash codes are integer values that identify objects.

Q.5.16. What is the Properties class?

Ans. The properties class is a subclass of Hashtable that can be read from or written to a stream.

It also provides the capability to specify a set of default values to be used if a specified key is not found in the table. We have two methods load() and save().

Q.5.17. Can null value be added to a List?

Ans. Yes.

A Null value may be added to any List.

Q.5.18. What is the ResourceBundle?

Ans. The ResourceBundle class also supports internationalization.

ResourceBundle subclasses are used to store locale-specific resources that can be loaded by a program to tailor the program's appearence to the paticular locale in which it is being run. Resource Bundles provide the capability to isolate a program's locale-specific resources in a standard and modular manner.

Q.5.19. How are Observer Interface and Observable class, in java.util package, used?

Ans. Objects that subclass the Observable class maintain a list of Observers. When an Observable object is updated it invokes the update() method of each of its observers to notify the observers that it has changed state. The Observer interface is implemented by objects that observe Observable objects.

Q.5.20. Explain in deep differences between an Abstract class and Interface ?

Ans.

Feature	Interface	Abstract class
multiple inheritance	A class may implement several interfaces.	A class may extend only one abstract class.
Default imple-mentation	An interface cannot provide any code at all, much less default code.	An abstract class can provide complete code, default code, and/or just stubs that have to be overridden.
constants	Static final constants only, can use them without qualification in classes that implement the interface. On the other paw, these unqualified names pollute the namespace. You can use them and it is not obvicus where they are coming from since the qualification is optional.	Both instance and static constants are possible. Both static and instance initialize code are also possible to compute the constants.
third party convenience	An interface implementation may be added to any existing third party class.	A third party class must be rewritten to extend only from the abstract class.
is-a vs -able or can-do	Interfaces are often used to describe the peripheral abilities of a class, not its central identity, e.g. an Automobile class might implement the Recyclable interface, which could apply to many otherwise totally unrelated objects.	An abstract class defines the core identity of its descendants. If you defined a Dog abstract class then Damamation descendants **are** Dogs, they are not merely dogable. Implemented interfaces enumerate the general things a class can do, not the things a class is. In a Java context, users should typically implement the Runnable interface rather than extending Thread, because they're not really interested in providing some new Thread functionality, they normally just want some code to have the capability of running independently. They want to create something that can be run in a thread, not a new kind of thread.The similar is-a vs has-a debate comes up when you decide to inherit or delegate.
plug-in	You can write a new replacement module for an interface that contains not one stick of code in common with the existing implementations. When you implement the inteface, you start from scratch without any default implementation. You have to obtain your tools from other classes; nothing comes with the interface other than a few constants. This gives you freedom to implement a radically different internal design.	You must use the abstract class as-is for the code base, with all its attendant baggage, good or bad. The abstract class author has imposed structure on you. Depending on the cleverness of the author of the abstract class, this may be good or bad.

Feature	Interface	abstract class
Homo-geneity	If all the various implementaions share is the method signatures, then an interface works best.	If the various implementations are all of a kind and share a common status and behaviour, usually an abstract class works best. Another issue that's important is what I call "heterogeneous vs. homogeneous." If implementors/ subclasses are homogeneous, tend towards an abstract base class. If they are heterogeneous, use an interface. (Now all I have to do is come up with a good definition of hetero/homo-geneous in this context.) If the various objects are all of-a-kind, and share a common state and behavior, then tend towards a common base class. If all they share is a set of method signatures, then tend towards an interface.
Maintenance	If your client code talks only in terms of an interface, you can easily change the concrete implementation behind it, using a factory method.	Just like an interface, if your client code talks only in terms of an abstract class, you can easily change the concrete implementation behind it, using a factory method.
Speed	Slow, requires extra indirection to find the corresponding method in the actual class. Modern JVMs are discovering ways to reduce this speed penalty.	Fast
terseness	The constant declarations in an interface are all presumed public static final, so you may leave that part out. You can't call any methods to compute the initial values of your constants. You need not declare individual methods of an interface abstract. They are all presumed so.	You can put shared code into an abstract class, where you cannot into an interface. If interfaces want to share code, you will have to write other bubblegum to arrange that. You may use methods to compute the initial values of your constants and variables, both instance and static. You must declare all the individual methods of an abstract class abstract.
adding functionality	If you add a new method to an interface, you must track down all implementations of that interface in the universe and provide them with a concrete implementation of that method.	If you add a new method to an abstract class, you have the option of providing a default implementation of it. Then all existing code will continue to work without change.

Q.5.21. Which java.util classes and interfaces support event handling?

Ans. The EventObject class and the EventListener interface support event processing.

Q.5.22. Does java provide standard iterator functions for inspecting a collection of objects?

Ans. The Enumeration interface in the java.util package provides a framework for stepping once through a collection of objects. We have two methods in that interface.

```
public interface Enumeration {
```

```
boolean hasMoreElements();
Object nextElement();
}
```

Q.5.23. The Math.random method is too limited for my needs- How can I generate random numbers more flexibly?

Ans. The random method in Math class provide quick, convienient access to random numbers, but more power and flexibility use the Random class in the java.util package.

```
double doubleval = Math.random();
```

The Random class provide methods returning float, int, double, and long values.

```
nextFloat() // type float; 0.0 <= value < 1.0
nextDouble() // type double; 0.0 <= value < 1.0
nextInt() // type int; Integer.MIN_VALUE <= value <= Integer.MAX_VALUE
nextLong() // type long; Long.MIN_VALUE <= value <= Long.MAX_VALUE
nextGaussian() // type double; has Gaussian("normal") distribution with mean 0.0 and standard deviation 1.0)
Eg. Random r = new Random();
float floatval = r.nextFloat();
```

Q.5.24. How can we get all public methods of an object dynamically?

Ans. By using getMethods(). It return an array of method objects corresponding to the public methods of this class.

getFields() returns an array of Filed objects corresponding to the public Fields(variables) of this class.

getConstructors() returns an array of constructor objects corresponding to the public constructors of this class.

Q.5.25. What is CLASSPATH?

Ans. CLASSPATH is an environment variable which tells the Java compiler and the JVM where to look for Java class files.

In Java 1.2 and later, the CLASSPATH variable is less critical because Java knows how to find the default classes and because Java loads classes from the current directory by default.

If you do not want to use an environment variable, you can use the -classpath argument to the Java compiler or the JVM as such:

```
javac -classpath . JavaProg
```

To add a jar file to your CLASSPATH, include the entire pathname of the jar file — including the filename.

EXERCISES

Q.5.1. Consider the following two classes:

```
public class classA
{
        public void methodOne (int i)
            {        }
        public void methodTwo (int i)
            {        }
        public static void methodThree (int i)
            {        }
```

```
public static void methodFour (int i)
    {          }
}
public class classB
{
        public static void methodOne (int i)
            {         }
        public void methodTwo (int i)
            {         }
        public void methodThree (int i)
            {         }
}
```

(*a*) which method overrides a method in the superclass?
(*b*) which method hides a method in the superclass?
(*c*) what do the other methods do?

Q.5.2. Explain Inheritance with types.

Q.5.3. What is the mean of derived class?

Q.5.4. What are the structures of different type of inheritance?

Q.5.5. What is the meant by method overriding?

Q.5.6. Can we override the constructors?

Q.5.7. What do you mean by abstract class and Abstract method and why we need it?

Q.5.8. What is the concept of package? What are the advantages of package?

Q.5.9. What do you mean by CLASSPATH?

Q.5.10. What do you mean by jar file?

Q.5.11. What is Interface? Explain.

Q.5.12. What are the difference between abstract class and interface?

Q.5.13. How can we define an Interface? Give an example.

❑❑❑

Chapter 6

Advance Programming Concept

CHAPTER OUTLINE

- Exception Handling
  - Exception Handling Fundamentals
  - Types of Exceptions
  - Divide-By-Zero Exception
  - Multiple catch Clause
  - Nested try Statements
  - throw and throws
  - finally
  - Built-In Exceptions
- Multithreaded Programming
  - Multitasking
  - Thread Vs Processes
  - What is thread?
  - The Thread Classes
  - A Simple Thread
  - The Life Cycle of a Thread
  - Thread Priorities
- Input/Output
  - Console Input
  - Integer input
  - Console Output
  - Streams
  - File Input
  - String Tokenizer
  - File Output

6.1. EXCEPTION HANDLING.

6.1.1. Exception Handling Fundamentals:

The term *exception* is shorthand for the phrase "exceptional event."

When an error occurs within a method, the method creates an object and hands it off to the runtime system. The object, called an *exception object*, contains information about the error, including its type and the state of the program when the error occurred. Creating an exception object and handing it to the runtime system is called *throwing an exception.*

After a method throws an exception, the runtime system attempts to find something to handle it. The set of possible "somethings" to handle the exception is the ordered list of methods that had been called to get to the method where the error occurred. The list of methods is known as the *call stack* (see figure 6.1).

The runtime system searches the call stack for a method that contains a block of code that can handle the exception. This block of code is called an *exception handler.* The search begins with the method in which the error occurred and proceeds through the call stack in the reverse order in which the methods were called. When an appropriate handler is found, the runtime system passes the exception to the handler.

An exception handler is considered appropriate if the type of the exception object thrown matches the type that can be handled by the handler.

> **Concept**
>
> *An exception is an event, which occurs during the execution of a program, that disrupts the normal flow of the program's instructions.*

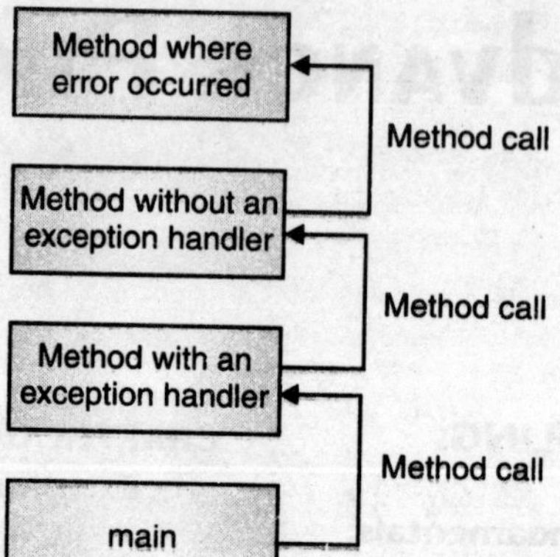

Fig. 6.1. The call stack

The exception handler chosen is said to *catch the exception.* If the runtime system exhaustively searches all the methods on the call stack without finding an appropriate exception handler, as shown in the figure 6.2 the runtime system (and, consequently, the program) terminates.

> **Concept**
>
> *An exception is a condition that is caused by a runtime error in the program.*

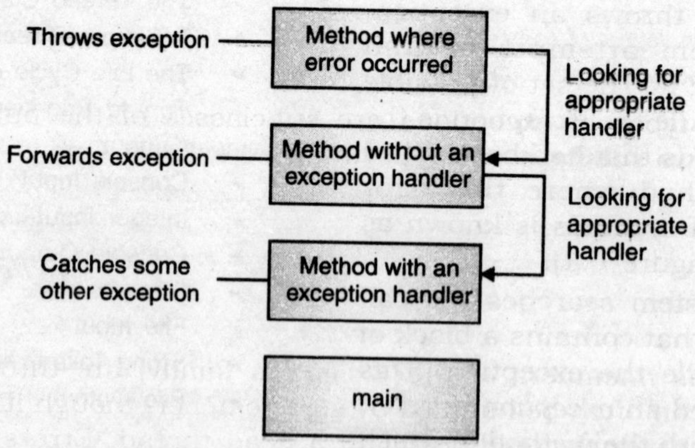

Fig. 6.2. Searching the call stack for the exception handler

A Java exception is an object that describes an exceptional (that is, an error) condition that has occurred in a piece of code. When an exceptional condition arises, an object representing that exception is created and thrown in the method that caused the error. That method may choose to handle that exception or pass it on. Either way, at some point, the exception is caught and processed.

Briefly, here is how they work. Program statements that should be monitored for exceptions are contained with in a **try block**. If an exception occurs within the try block, it is **thrown**. Our code can catch this exception (using **catch**) and handle it in some rational manner. System generated exceptions are automatically thrown by the Java run-time system. To throw an exception manually, use the keyword **throw**. An exception that is thrown out of a method must be specified as such by a throws clause. Any code that absolutely must be executed before a method returns is put in a **finally block**.

This is the general form of an exception-handling block:

```
try
{
      //block of code to be monitored for errors
}

catch (ExceptionType1 exOb )
{
      //exception handler for ExceptionType1
}
catch (ExceptionType2 exOb )
{
//exception handler for ExceptionType2
}

//...
finally
{
//block of code to be executed before try block ends
}
```

6.1.2. Types of Exceptions

In Java, all types of exceptions are subclasses of the built-in class **Throwable**. **Throwable** has two subclasses that partition exceptions. One is Exception and the other is Error. There is an important subclass of Exception called **RuntimeException** .

Throwable serves as the base class for an entire family of classes, declared in java.lang, that your program can instantiate and throw. A small part of this family is shown in figure 6.3.

As you can see in figure 6.3, Throwable has two direct subclasses, Exception and Error. Exceptions (members of the Exception family) are thrown to signal abnormal conditions that can often be handled by some catcher, though it's possible they may not be caught and therefore could result in a dead thread. Errors (members of the Error family) are usually thrown for more serious problems, such as OutOfMemoryError, that may not be so easy to handle. In general, code you write should throw only exceptions, not errors. Errors are usually thrown by the methods of the Java API, or by the Java virtual machine itself.

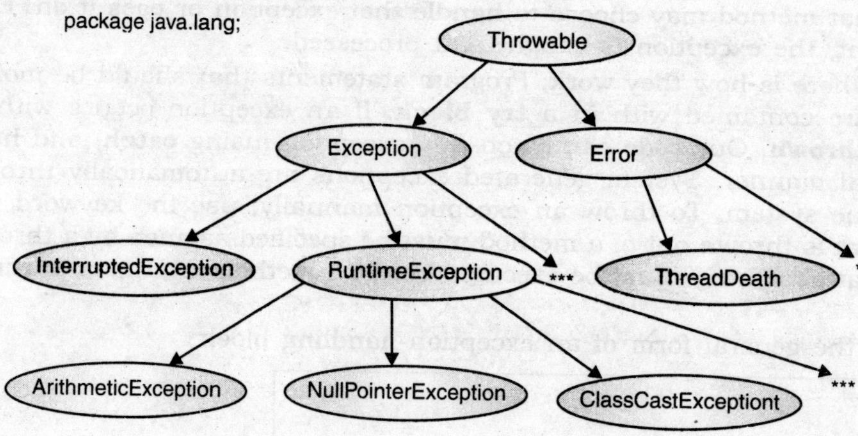

Fig. 6.3. A Partical view of the Throwable family

In addition to throwing objects whose classes are declared in java.lang, you can throw objects of your own design. To create your own class of throwable objects, you need only declare it as a subclass of some member of the Throwable family. In general, however, the throwable classes you define should extend class Exception. They should be "exceptions."

6.1.3. DivideBy Zero Exception

This small program has an expression that causes a divide-by-zero exception.

```
class DivideByZero
{
    public static void main (String args[])
    {
        int d = 0;
        int a = 42 / d;
    }
}
```

The Java run-time system constructs a new exception object when it detects an attempt to divide-by-zero. It then throws this exception. In the above example, there is no exceptional handler to catch the exception. The default handler provided by the Java run-time system will process any exception that is not caught by a program. The output of the above program when executed by the standard Java JDK runtime interpreter:

```
java.lang.ArithemeticException:/ by zero
at DivideByZero.main(DivideByZero.java:4)
```

Although the default exception handler is provided by Java run-time system, you will usually like to handle exception yourself. The following program includes a try block and a catch clause, which processes the ArithmeticException generated by the division-by-zero:

Example 6.1: DivideByZero Exception

```
class DivideByZero
{
```

```
public static void main (String args[])
{
int d, a;
try //monitor a block of code
{
d = 0;
a = 42 / d;
System.out.println("This will not be printed. ");
}
catch(ArithmeticException e)
{
System.out.println("Division by zero.");
}
System.out.println("After catch statement.");
}
```

OUTPUT:

```
Division by zero.
After catch statement
```

The call to println() inside the try block is never executed .Once an exception is thrown, program control transfers out of the try block into the catch block. Once the catch statement has executed, the program continues with the next line in the program following the entire try/catch mechanism.

6.1.4. Multiple Catch Clause

In some situations, we need to maintain more than one catch clause for a block of code. In such situations when an exception is thrown, each catch statement is inspected in order and the first one whose type matches that of the exception is executed. After one catch statement executes, the others are bypassed, and execution continues after the try/catch block.

Example 6.2: Multiple catch statements.

```
class Multicatch
{
public static void main (String args[])
{
try
{
int a = args.length;
System.out.println("a = " + a);
int b = 42 / a;
int c[ ] = { 1 };
c[42] = 99;
}
catch (ArithmeticException e)
{
    System.out.println("Divide by 0 " );
```

```
    }
    catch (ArrayIndexOutOfBoundException e)
    {
        System.out.println("Array index is out of bounds");
    }
        System.out.println("After try/catch blocks");
    }
}
```

SAMPLE OUTPUTS:

```
1. C:\> java Multicatch
a = 0
Divide by 0
After try/catch blocks
2. C:\> java Multicatch TestArg
a = 1
Array index is out of bounds
After try/catch blocks
```

When we use multiple catch statements, it is important to remember that exception subclasses must come before any of their superclasses. This is because a catch statement that uses a superclass will catch exceptions of that type plus any of its subclasses. Thus, a subclass would never be reached if it came after its superclass. Further, in Java unreachable code is an error.

6.1.5. Nested try Statements

In Java, a try statement can be inside the block of another try. Each time a try statement is entered, the context of that exception is pushed on the stack .If an inner try statement does not have a matching catch statement for a particular exception, the stack is unwound and the next try statement's catch handlers are inspected for a match. This continues until one of the catch statements succeeds, or until all of the nested try statements are exhausted. If no catch statements match, then the Java run-time system will handle the exception.

Example 6.3: Nested try statements.

```
class NestTry
{
 public static void main (String args[])
 {
 try
 {
        int a = args.length;
        /* if no command-line args are present
        the following statement will generate
        a divide-by-zero exception*/
        int b = 42 / a;
        System.out.println("a = " + a);
        try //nested try block
```

```
        {
        /* if one command line arg is used,then a divide-by-zero exception
        will be generated by the following code*/
        if (a = = 1) a = a / (a - a); // division by zero

        /* if two command line arg is used ,
        then generate an out of bounds exception .*/
        if( a = 2 )
            {
            int c [] = { 1 };
            c[ 42 ] = 99; //generate an out of bound exception
            }
        }
    catch (ArrayIndexOutOfBoundsException e)
    {
        System.out.println("Array index out of bounds ");
    }
    }
    catch (ArithmeticException e)
    {
    System.out.println("Divide by zero");
    }
  }
  }
```

SAMPLE OUTPUTS:

```
1. C:\> java NestTry
Divide by zero.
2. C:\> java NestTry One
a = 1
Divide by zero
3)C:\> java NestTry One Two
a = 2
Array index out of bounds.
```

6.1.6. Throw and Throws

So far, we have been catching exceptions that are thrown by the Java run-time system

It is possible for our program to throw an exception explicitly using **throw** statement. The usual form of throw is shown here.

```
throw throwableinstance;
```

Here *throwableinstance* must be an object of type **Throwable** or subclass of **Throwable** .The flow of execution stops immediately after the throw statement any subsequent statements are not executed. The nearest enclosing try block is inspected to see if it has a catch statement that matches the type of exception. If it does find a match, control is transferred to that statement. If not then the next enclosed try statement is inspected, and so on......If no matching catch is found, then the default exception handler halts the program.

Here the a sample program

Example 6.4: Demonstrate throw.

```
class ThrowDemo {
  static void demoproc() {
  try {
        throw new NullPointerException("demo");
  }
  catch(NullPointerException e) {
  System.out.println("Caught inside demoproc.");
  throw e; // re-throw the exception
  }
 }

 public static void main(String args[]) {
  try {
        demoproc();
  }
  catch(NullPointerException e) {
  System.out.println("Recaught " );
  }
 }
}
```

OUTPUT

```
Caught inside demoproc.
Recaught
```

Here, **new** is used to construct an instance of **NullPointerException**. All of Java's built-in run-time exceptions have two constructors: one with no parameter, and one that takes a string parameter that describes the exception.

If a method is capable of causing an exception that it does not handle, it must specify this behavior so that callers of that method can guard themselves against that exception. We can do this by including a **throws** class in the methods declaration as follows.

```
type method-name(parameter-list) throws exception-list
{
//Body of the method.
}
```

exception-list is a list of types of exceptions that a method might throw . All exceptions except those of type **Error** or **RuntimeException**, or any of their subclasses need not be listed.

Example 6.5: Throws

```
class ThrowsDemo
{
```

```
static void throwOne() throws IllegalAccessException
    {
    System.out.println("Inside throwOne.");
    throw new IllegalAccessException("demo");
    }
public static void main(String args[])
{
try
{
    throwOne();
}
catch (IllegalAccessException e)
{
    System.out.println("Caught ");
    }
}
}
```

OUTPUT

```
Inside throwOne.
Caught
```

6.1.7. Finally

Finally creates a block of code that will be executed after the try/catch block has completed and before the code following the try/catch block. The **finally** block will execute whether or not an exception is thrown, even if no catch statement matches the exception. Here is an example:

Example 6.6: Demonstrate finally.

```
class FinallyDemo
{
 // Through an exception out of the method.
 static void procA()
 {
     try
     {
     System.out.println("inside procA");
     throw new RuntimeException("demo");
     }
     finally
     {
     System.out.println("procA's finally");
     }
 }

 // Return from within a try block.
 static void procB()
 {
```

```
        try
            {
            System.out.println("inside procB");
            return;
            }
    finally
    {
        System.out.println("procB's finally");
    }
    }

    // Execute a try block normally.
    static void procC()
    {
        try
        {
        System.out.println("inside procC");
            }
    finally
        {
        System.out.println("procC's finally");
        }
    }

    public static void main(String args[])
    {
        try
        {
            procA();
        }
        catch (Exception e)
        {
            System.out.println("Exception caught");
        }
        procB();
        procC();
    }
}
```

OUTPUT:

```
inside procA
procA's finally
Exception caught
inside procB
procB's finally
inside procC
procC's finally
```

6.1.8. Built-in Exceptions

Inside the standard package **java.lang** , Java defines several exception classes. In Java there are some exceptions (exceptions derived from **RuntimeException**) which compiler does not check to see if a method throws or handles these exceptions. Such exceptions are called unchecked exceptions. All the exceptions defined by java.lang that must be included in method's throws list are called checked exceptions.

Unchecked RuntimeException Subclasses

Exception	Meaning
ArithmeticException	Arithmetic error, such as divide by zero.
ArrayIndexOutOfBoundsException	Array index is out of bounds.
ArrayStoreException	Assignment to an array element of an in compatible type
ClassCastException	Invalid cast
IllegalArgumentException	Illegal argument used to invoke a method
IllegalMonitorStateException	Illegal monitor operation, such as waiting on an unlocked thread.
IllegalStateException	Environment or application is in incorrect state
IllegalThreadStateException	Requested operation not compatible with current thread state
IndexOutOfBoundsException	Some type of Index is out of bounds
NegativeArraySizeExeption	Array created with the negative size
NullpointerException	Invalid use of null exception
NumberFormatException	Invalid conversion of a string to a numeric format
SecurityException	Attempt to violate security
StringIndexOutOfBounds	Attempt to index outside the bounds of a string
UnsupportedOperationException	An unsupported operation was encountered

Checked Exceptions Defined in java.lang

Exception	Meaning
ClassnotFoundException	Class not found
CloneNotSupportedException	Attempt to clone an object that does not implement the cloneable interface
IllegalAccessException	Access to a class is denied
InstantiationException	Attempt to create an Object of an abstract class or interface
InterruptedException	One thread has been interrupted by another thread
NoSuchFieldException	A requested field does not exist
NoSuchMethodException	A requested method does not exist

6.2. MULTITHREADED PROGRAMMING

6.2.1. Multitasking

The earliest computers did one thing at a time. A lab computer might calculate detonation waves for one of the physicists. Next it would calculate mixing ratios for a chemist. All programs were run sequentially, one at a time; and each had full control of the computer. However two programs couldn't run at the same time. This is called batch processing, and it's a very efficient way to get maximum usage out of a very expensive computer. Almost all the computer's time is spent actually calculating jobs and not switching between one program and another. However batch processing can be very annoying when your differential equation integration program that would take two seconds of CPU time gets stuck in line behind the physics department's nuclear structure modeling project that's going run for the next three days.

Time sharing operating systems were invented to allow multiple people to use one then very expensive computer at the same time. On a time sharing system many people can run programs at the same time. The operating system is responsible for splitting the time among the different programs that are running. That way you can finish integrating your differential equation while the physics department's nuclear modeling program is still churning away. The physics department nuclear modeling program might take two weeks to run instead of three days, but everyone with the shorter programs is happy (at least until the physicists figured out how to hack the computer so that it only ran their program).

> **Concept**
>
> *The multitasking is the ability of single processor to perform more than operation at the same time.*

Once systems allowed different users to run programs at the same time it was a short step to letting the same user run multiple programs simultaneously. Each running program (generally called a *process*) had its own memory space, its own set of variables, its own stack and heap, and so on. One process could produce or call another process, but afterwards the two processes behaved more or less independently. Mechanisms like inter procedure calls (IPC) were developed to allow processes to interact with each other, but such interaction was expensive and difficult. And this is where matters rested for about twenty years.

However it's not just users that want to do different things at the same time. Many programs also need to do several things at once. A web browser, for instance, can print a file in the background while it downloads a page in one window and formats the page as it downloads. The ability of an individual program to do more than one thing at the same time is most efficiently implemented through threads.

6.2.2. Thread Vs Processes

Both threads and processes are methods of parallelizing an application. However, processes are independent execution units that contain their own state information, use their own address spaces, and only interact with each other via interprocess communication mechanisms (generally managed by the operating system). Applications are typically divided into processes during the design phase, and a master process explicitly call sub-processes when it makes sense to logically separate significant application functionality. Processes, in other words, are an architectural construct.

> **Concept**
>
> *A single process may contain more than one threads.*

By contrast, a thread is a coding construct that doesn't affect the architecture of an application. A single process might contains multiple threads; all threads within a process share the same state and same memory space, and can communicate with each other directly, because they share the same variables.

Threads typically are called/invoked for a short-term benefit that is usually visualized as a serial task, but which doesn't have to be performed in a linear manner (such as performing a complex mathematical computation using parallelism, or initializing a large matrix), and then are absorbed when no longer required. The scope of a thread is within a specific code module—which is why we can bolt-on threading without affecting the broader application.

6.2.3. What is thread?

A *thread* can be loosely defined as a separate stream of execution that takes place simultaneously with and independently of everything else that might be happening. A thread is like a classic program that starts at point A and executes until it reaches point B. It does not have an event loop. A thread runs independently of anything else happening in the computer. Without threads an entire program can be held up by one CPU intensive task or one infinite loop, intentional or otherwise. With threads the other tasks that don't get stuck in the loop can continue processing without waiting for the stuck task to finish.

> **Concept**
>
> *A threads is a light-weighted process which runs concurrently with other threads. All threads of program defines a seprate path of execution.*

It turns out that implementing threading is harder than implementing multitasking in an operating system. The reason it's relatively easy to implement multitasking is that individual programs are isolated from each other. Individual threads, however, are not. To return to the printing example, suppose that while the printing is happening in one thread, the user deletes a large chunk of text in another thread. What's printed? The document as it was before the deletion? The document as it was after the deletion? The document with some but not all of the deleted text? Or does the whole system go down in flames? Most often in a non-threaded or poorly threaded system it's the latter.

Threaded environments like Java allow a thread to put locks on shared resources so that while one thread is using data no other thread can touch that data. This is done with *synchronization*. Synchronization should be used sparingly since the purpose of threading is defeated if the entire system gets stopped waiting for a lock to be released. The proper choice of objects and methods to synchronize is one of the more difficult things to learn about threaded programming.

6.2.4. The Thread Classes

Java has two ways a program can implement threading. One is to create a subclass of java.lang.Thread. However sometimes you'll want to thread an object that's already a subclass of another class. Then you use the java.lang.Runnable interface.

The Thread class has three primary methods that are used to control a thread:

```
public void start()
public void run()
public final void stop()
```

The start() method prepares a thread to be run; the run() method actually performs the work of the thread; and the stop() method halts the thread. The thread *dies* when the run() method terminates or when the thread's stop() method is invoked.

You never call run() explicitly. It is called automatically by the runtime as necessary once you've called start(). There are also methods to **suspend** and **resume** threads, to put threads to sleep and wake them up, to yield control to other threads, and many more we'll discuss these later.

The Runnable interface allows you to add threading to a class which, for one reason or another, cannot conveniently extend Thread. It declares a single method, run():

```
public abstract void run()
```

By passing an object which implements Runnable to a Thread() constructor, you can substitute the Runnable's run() method for the Thread's own run() method. (More properly the Thread object's run() method simply calls the Runnable's run() method.)

6.2.5. A Simple Thread

When writing a threaded program you can pretend that you're writing many different programs, each with its own run() method. Each thread is a subclass of java.lang.Thread. The following program is a thread that prints the numbers between -128 and 127.

```
public class BytePrinter extends Thread {
public void run() {
 for (int b = -128; b < 128; b++) {
 System.out.println(b);
 }
 }
}
```

You launch this thread from another method, probably in another class, by instantiating an object of this class using new and calling its start() method. To create a thread just call the default constructor for your subclass of Thread. For instance

```
BytePrinter bp = new BytePrinter();
```

This class only has the default, no arguments constructor; but there's absolutely no reason you can't add other constructors to your Thread subclass.

Constructing a thread object puts it at the starting line. The bell goes off and the thread starts running when you call the thread's start() method like this:

```
bp.start();
```

Once the start() method is called program execution splits in two. Some CPU time goes into whatever statements follow bp.start() and some goes into the bp thread. It is unpredictable which statements will run first. Most likely they will be intermixed. The bp thread will now continue running until one of seven things happens

- bp's run() method completes.
- bp's stop() method is called.
- bp's suspend() method is called.
- bp's sleep() method is called.
- bp's yield() method is called.
- bp blocks waiting for an unavailable resource
- Another thread preempts this thread.

Once program control reaches the end of bp's run() method the thread dies. It cannot be restarted, though you can create a new instance of the same Thread subclass and start that.

For example, the following program launches a single BytePrinter thread:

Eaxmple 6.7: Simple Thread

```
public class ThreadTest
{
 public static void main(String[] args)
 {

 System.out.println("Constructing the thread...");
 BytePrinter bp = new BytePrinter();
 System.out.println("Starting the thread...");
 bp.start();
 System.out.println("The thread has been started.");
 System.out.println("The main() method is finishing.");
 return;

 }

}
```

OUTPUT:

```
Constructing the thread...
Starting the thread...
The thread has been started.
The main() method is finishing.
-128
-127
-126
-125
-124
-123
-122
-121
-120
...
Of course this continues for a couple hundred more lines.
```

The following program launches three BytePrinter threads:

Example 6.8: Multiple threads

```
public class ThreadsTest
{

 public static void main(String[] args)
 {

 BytePrinter bp1 = new BytePrinter();
 BytePrinter bp2 = new BytePrinter();
 BytePrinter bp3 = new BytePrinter();
 bp1.start();
 bp2.start();
 bp3.start();

 }

}
```

The order of the output you see from this program is implementation dependent, and mostly unpredictable. It may look something like this:

```
-128
-127
-126
...
125
126
127
-128
-127
-126
...
125
126
127
-128
-127
-126
...
125
126
127
```

In this case the three threads run sequentially, one after the other. However on a few systems you may see something different

However on a few systems you may see something like this:

```
-128
-127
```

```
-126
-125
-124
-123
-128
-127
-126
-125
-128
-127
-126
-125
-124
-123
-122
-121
-120
-119
-124
-123
...
125
126
127
```

In this case output from the three different threads is intermixed. Some systems use cooperative threads that require a thread to explicitly yield control before other, equal priority threads get a chance to run. Other systems use preemptive threading in which the virtual machine guarantees that all threads of the same priority get time in which to run. However, even on preemptive VMs this thread takes so little time to run it's unlikely any intermixing will occur.

6.2.6. The Life Cycle of a Thread

Now that you've seen how to give a thread something to do, we'll review some details that were glossed over in the previous section. In particular, we look at the life cycle of a thread: how to create and start a thread, some of the special things it can do while it's running, and how to stop it.

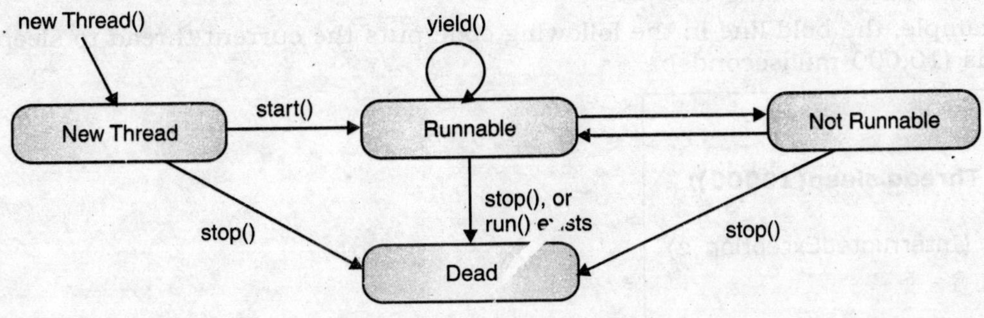

Fig. 6.4. Life cycle of thread

The following figure 6.4 illustrates the various states that a Java thread can be in at any point during its life. It also illustrates which method calls cause a transition to another state. This diagram is not a complete finite state diagram, but rather an overview of the more interesting and common facets of a thread's life. The remainder of this page discusses a Thread's life cycle in terms of its state.

New Thread

The following statement creates a new thread but does not start it, thereby leaving the thread in the "New Thread" state.

```
Thread myThread = new MyThreadClass();
```

When a thread is in the "New Thread" state, it is merely an empty Thread object. No system resources have been allocated for it yet. Thus when a thread is in this state, you can only start the thread or stop it. Calling any method besides start() or stop() when a thread is in this state makes no sense and causes an IllegalThreadStateException.

Runnable

Now consider these two lines of code:

```
Thread myThread = new MyThreadClass();
myThread.start();
```

The start() method creates the system resources necessary to run the thread, schedules the thread to run, and calls the thread's run() method. At this point the thread is in the "Runnable" state. This state is called "Runnable" rather than "Running" because the thread might not actually be running when it is in this state. Many computers have a single processor, making it impossible to run all "Runnable" threads at the same time. So, the Java runtime system must implement a scheduling scheme that shares the processor between all "Runnable" threads. (See Thread Priority for more information) However, for most purposes you can think of the "Runnable" state as simply "Running". When a thread is running—it's "Runnable" and is the current thread—the instructions in its run() method are executing sequentially.

Not Runnable

A thread enters the "Not Runnable" state when one of these four events occurs:

- Someone invokes its sleep() method.
- Someone invokes its suspend() method.
- The thread uses its wait() method to wait on a condition variable.
- The thread is blocking on I/O.

For example, the bold line in the following code puts the current thread to sleep for 10 seconds (10,000 milliseconds):

```
try
{
        Thread.sleep(10000);
}
catch (InterruptedException e)
{
}
```

During the 10 seconds that myThread is asleep, even if the processor becomes available myThread does not run. After the 10 seconds are up, myThread becomes "Runnable" again and, if the processor becomes available, runs.

For each of the entrances into the "Not Runnable" state shown in the figure 6.4, there is a specific and distinct escape route that returns the thread to the "Runnable" state. An escape route only works for its corresponding entrance. For example, if a thread has been put to sleep, then the specified number of milliseconds must elapse before the thread becomes "Runnable" again. Calling resume() on a sleeping thread has no effect.

The following indicates the escape route for every entrance into the "Not Runnable" state.

- If a thread has been put to sleep, then the specified number of milliseconds must elapse.
- If a thread has been suspended, then someone must call its resume() method.
- If a thread is waiting on a condition variable, whatever object owns the variable must relinquish it by calling either notify() or notifyAll().
- If a thread is blocked on I/O, then the I/O must complete.

Dead

A thread can die in two ways: either from natural causes, or by being killed (stopped). A thread dies naturally when its run() method exits normally. For example, the while loop in this method is a finite loop—it will iterate 100 times and then exit.

```
public void run()
{
    int i = 0;
    while (i < 100)
    {
        i++;
        System.out.println("i = " + i);
    }
}
```

A thread with this run() method will die naturally after the loop and the run() method completes.

You can also kill a thread at any time by calling its stop() method. The following code snippet creates and starts myThread then puts the current thread to sleep for 10 seconds. When the current thread wakes up, the bold line in the code segment kills myThread.

```
Thread myThread = new MyThreadClass();
myThread.start();
try
{
    Thread.sleep(10000);
}
catch (InterruptedException e)
{
}
myThread.stop();
```

The stop() method throws a ThreadDeath object at the thread to kill it. Thus when a thread is killed in this manner it dies asynchronously. The thread will die when it actually receives the **ThreadDeath** exception.

The stop() method causes a sudden termination of a Thread's run() method. If the run() method performs critical or sensitive calculations, stop() may leave the program in an inconsistent or awkward state. Normally, you should not call Thread's stop() method but arrange for a gentler termination such as setting a flag to indicate to the run() method that it should exit.

IllegalThreadStateException

The runtime system throws an IllegalThreadStateException when you call a method on a thread and that thread's state does not allow for that method call. For example, **IllegalThreadStateException** is thrown when you invoke suspend() on a thread that is not "Runnable".

As shown in the various examples of threads so far in this chapter, when you call a thread method that can throw an exception, you must either catch and handle the exception, or specify that the calling method throws the uncaught exception.

The isAlive() Method

A final word about thread state: the programming interface for the Thread class includes a method called isAlive(). The isAlive() method returns true if the thread has been started and not stopped. Thus, if the isAlive() method returns **false** you know that the thread is either a "New Thread" or "Dead". If the isAlive() method returns **true**, you know that the thread is either "Runnable" or "Not Runnable". You cannot differentiate between a "New Thread" and a "Dead" thread; nor can you differentiate between a "Runnable" thread and a "Not Runnable" thread.

6.2.7. Thread Priorities

The scheduler determines which thread should be running based on a priority number assigned to each thread. The range of priorities is from 1 to 10. The default priority of a thread is Thread.NORM_PRIORITY, which is assigned the value of 6. Two other static variables are made available, they are Thread.MIN_PRIORITY, which is set to 1, and Thread.MAX_PRIORITY, which is set to 10. The getPriority() method can be used to find the current value of the priority of a thread.

6.3. JAVA INPUT/OUTPUT

Input is any information that is needed by your program to complete its execution. There are many forms that program input may take. Some programs use graphical components like a popup dialog box to accept and return the character string that is typed by the user. You are certainly familiar with programs that are controlled simply by clicking the mouse in a specific area of the screen. Still other programs, like word processing programs, get some of their input from a file that is stored on the computer's floppy or hard disk drive. Some programs, like web browsers, get their data from a network connection, while others get data from devices like scanners, digital cameras and microphones. The possibilities are limited only by computer scientists' imagination.

Output is any information that the program must convey to the user. The information you see on your computer screen is being output by one or more programs that are currently running on your computer. When you decide to print a document, a program is told to send some output to the printer. Any sound that your computer makes is because some program sent output to the speakers on your computer. The possibilities for program output are also limited only by our imaginations.

There are several potential error conditions that may occur when a program needs to get input from the user. If a user enters letters when a program is expecting numbers, an exception [error] will occur if the program assumes incorrectly that it has a valid integer to use in calculations. Programs must be written to survive bad input by the user. One way to do this is to ensure that only valid input is accepted.

Standard Java classes do not ensure that only valid input is accepted. They are designed to be very flexible to support the wide variety of input and output options available now and in the future. This flexibility comes at the cost of increased complexity.

6.3.1. Console Input

The console window is the [black] window that is automatically launched when you run a program. Console input is any input that is entered in the console window instead of typing it into a field or dialog box that pops up in a window.

When a program is waiting for input at the console, there is sometimes a blinking cursor in the console window indicating that the user should type some information. But, this is not always the case. The user will only know what information to type, if the program describes that information in the form of a user prompt.

The use of several of the Java I/O classes are required to successfully receive input that is typed by the user. The java.io package contains most, if not all, of the classes you will need to use. Don't worry, you won't need to use all 50+ classes. But, you will need to learn about and use at least three of them. All three classes are in the java.io package. Either use the fully qualified name shown or import the java.io package.

- **java.io.InputStream** - stores information about the connection between an input device and the computer or program.
- **java.io.InputStreamReader** - used to translate data bytes received from **InputStream** objects into a stream of characters.
- **java.io.BufferedReader** - used to buffer (store) the input received from a **InputStreamReader** object. This is done to improve the efficiency. Input devices are much slower than the computer's processor and buffering the data reduces the number of times the CPU has to interact with the device itself.

None of these classes has a method as convenient and as error proof as you may wish. However, the **BufferedReader** class does have a method called ***readLine*** that does return a line of text as typed by the user. There are two available constructors for creating a **BufferedReader** object. For console input, we will use the one that requires only one argument, an instance of a **Reader** object. That means we need to create an instance of the class **java.io.Reader**.

The **InputStreamReader** class *extends* the **Reader** class. Here's an analogy to explain *extends*: All Robins are Birds, but not all Birds are Robins. Therefore, Robin extends Bird. If someone needs a Bird, then a Robin can be used. This means that any instance of the **InputStreamReader** class can be used whenever an instance of the **Reader** class is required. The inheritance hierarchy of the **InputStreamReader** class shows that it *extends* the **Reader** class because the **Reader** class is higher in the hierarchy tree.

We will create an instance of the **InputStreamReader** class to be used to create the **BufferedReader** object that we want. We choose to create an **InputStreamReader** instead of a **Reader** object because we want to get input from an **InputStream**. Later, you will see that File I/O will require the use of a different type of **Reader** object.

What do we need to have in order to create an instance of the **InputStreamReader** class? According to the Java API, we will need an **InputStream** object. The **System** class in the java.lang package automatically creates an **InputStream** object that is connected to the keyboard. It is called System.in and is part of the java.lang package.

We will use the System.in object to create an instance of the **InputStreamReader** class and then use that object to create an instance of the **BufferedReader** class.

Steps for console based user input:

1. Use the System.in object to create an **InputStreamReader** object.
2. Use the **InputStreamReader** object to create a **BufferedReader** object.
3. Display a prompt to the user for the desired data.
4. Use the **BufferedReader** object to read a line of text from the user.
5. Do something interesting with the input received from the user.

The **BufferedReader** (and other I/O classes) are not in the standard java.lang package. You must import the java.io package to declare and create instances of any of the Java I/O classes. Add the import java.io.*; statement to your list of other import statements.

You will also have to inform the compiler that you are calling a method that may cause a **checked exception** to occur. Add the phrase *throws IOException* clause to the header of any method that calls stdin.readLine(). You will also need to add this clause to any method that calls your method that calls ***readLine***. Here's a complete program example that prompts the user for input and then repeats that data to the console window:

Example 6.9: Console based input and output.

```
import java.io.*; // needed for BufferedReader, InputStreamReader, etc.

public class MyConsoleIO
{
    // Create a single shared BufferedReader for keyboard input
    private static BufferedReader stdin =
    new BufferedReader( new InputStreamReader( System.in ) );

    public static void main ( String [] args ) throws IOException
    {
        System.out.print( "Type some data for the program: " );

        // Read a line of text from the user.
        String input = stdin.readLine();

        System.out.println( "input = " + input );

    }

}
```

6.3.2. Integer input

Getting data from the user isn't so hard after all. But, it does require some additional work. There is even more work to do, if you want to get an integer (or other numeric value) from the user. If the user types in "123", that will be still be returned as a **String** object by the ***readLine*** method of **BufferedReader**. You will need to *parse* [convert] the **String** object into an int value if you wish to store it in an int variable or data member. Here's how:

1. Get a **String** of characters that is in an integer format, eg. "123".

```
String input = stdin.readLine();
// from console input example above
```

2. Use the **Integer** class to parse the string of characters into an integer.

```
int number = Integer.parseInt( input );
// converts a String into an int value
```

The ***Integer*** class contains conversion methods for changing String data into int values and vice versa. The Integer class is one of several wrapper classes that are defined in the standard Java API. Wrapper classes have class methods for parsing and are also used when you need to store a primitive value as an object.

6.3.3. Console Output

We have used System.out.print(...) and System.out.println(...) statements for displaying simple text messages to the user. This is an important output alternative, since graphic user interface (GUI) objects are not readily available in some programming environments.

Most programming languages have the ability to display a string of characters to the screen or some other standard display device. We call this console output because the string of characters appears in a console window. The System.out object is an instance of the PrintStream class, which is a type of Stream.

6.3.4. Streams

A Stream object is used to store information needed to connect a computer program to an input or output device. Just like there is a Reader object that adds functionality to input streams, there is a Printer object that adds functionality to output streams. The PrintStream class *extends* the **Printer** class and contains definitions for all of the versions of the ***print*** and ***println*** methods that we use to display information like user prompts or results of calculations, etc.

Console output in Java is very easy because the ***print*** and ***println*** methods will work with any type of data. There is a separate version of each of these methods in the **PrintStream** class so that this is possible. There is also a version of the ***print*** and ***println*** methods that will print information for any object. But, how did we get a **PrintStream** object in the first place?

The **java.lang.System** class creates three different I/O streams automatically for us when our application begins execution. Each of these streams is public and static so that we can access them directly without having to create an instance of the **System** class. We have already used the **InputStream** object named System.in in the discussion on console input. The other two stream objects are named System.out and System.err. Each of these objects is an instance of the **PrintStream** class and is available for use in displaying information to the computer screen.

For example, if the following variables are defined,

```
int x = 3;
double rate = 6.5;
boolean playing = true;
String phrase = "The winner is ";
```

they can all be printed using ***print*** or ***println*** as follows:

```
System.out.print( "x = " + x + " rate = " );
System.out.println( rate );
System.out.println( "playing = " + playing );
System.out.println( phrase + "Deb" );
```

We can also print other types of data, including other objects, using the **print** and **println** methods. The following code fragment shows the command syntax for printing a **Wanderer** object. The class name **Wanderer** is used as an example. You can replace **Wanderer** with any class name that is defined in your program.

```
Wanderer wanderer1 = new Wanderer( "Wilma", Color.orange );
System.out.println( wanderer1 );
```

In this case, the program prints out some cryptic information about the **Wanderer** object. It is the class name, an @ symbol and the hexidecimal representation of the hashcode. The output looks like the following for the first Wanderer object .

```
Wanderer@13fac
```

Each object created has its own hashcode that can be used to distinguish it from other objects. However, hashcodes are not very readable for most users, so there is a way for the programmer to redefine what information is printed. The information that is displayed when an object is printed using the print method is defined by an instance method named ***toString***. Every class has a version of the ***toString*** method already defined that returns the information as described above. All classes inherit this method from the java.lang.Object class.

To redefine the ***toString*** method, you *override* the default version by defining a method with the same visibility and *method signature* as the inherited version of the method. The ***toString*** method of my **Wanderer** class can be *overridden* as follows:

```
public String toString()
{
    String coords = "(" + myLoc.getX() + "," + myLoc.getY() + ")";

    return myName + " is at " + coords;
}
```

Now, when the ***print*** or ***println*** method is used to print a **Wanderer** object, the new version of the ***toString*** method will be called instead of the version defined in the **Object** class. The **String** that is printed by the ***println*** method will look something like this:

```
Wilma is at (11,3)
```

Each class can and should override the ***toString*** method to return a **String** of characters that is more descriptive of the object than the default version provided by the **Object** class. This method can be called by any method that needs a **String** that describes the object.

The ***print*** or ***println*** methods of the **PrintStream** class should be adequate to produce most screen output. The only other type of output we will cover in file output. Other output devices require more specialized output objects and won't be covered in this book.

6.3.5. File Input

As mentioned above, data can be *read* from a variety of different sources, including data files stored on devices such as hard disk drives and floppy drives. The file will need to be opened and a **BufferedReader** will be *attached* to the file object. The process is actually very similar to the console input example above. The difference is that the **BufferedReader** will be created from a **FileReader** object instead of an **InputStreamReader** object.

The discussion and examples in this section explain the procedure when the file to be read is a text file that has valid ASCII characters to represent the data. Here are two new Java I/O classes to review:

- **java.io.File** - stores information about a file on a computer drive.
- **java.io.FileReader** - used to translate data bytes from File objects into a stream of characters.

Here's a code fragment to illustrate reading a text file. The ***readLine*** method may cause an **IOException** which is a checked exception, so be sure to *catch* the exception or add the *throws IOException* clause to the method header.

```
System.out.print( "Enter the filename: " );
String fileName = stdin.readLine(); // get a file name from the user

java.io.File file = new java.io.new File( fileName );
// create a File object

if ( file.exists() ) // check that the file exists
{ // before trying to create a BufferedReader
// Create a BufferedReader from the file
java.io.BufferedReader inFile = new java.io.BufferedReader(
new java.io.FileReader( file ) );

int lineNum = 0;

// Compare the results of calling the readLine method to null
// to determine if you are at the end of the file.

String line = inFile.readLine();
while ( line != null )
{
System.out.println( ++lineNum + ": " + line );
line = inFile.readLine();
}
```

```
// Close the buffered reader input stream attached to the file
inFile.close();
}
```

File Input Tips:

- Don't forget to import java.io.* classes.
- Add the *throws IOException* clause to any method that calls the ***readLine*** method of a **BufferedReader** object.
- Do not assume that the ***readLine*** will return valid data. You should test the return value of calling the ***readLine*** method against null before trying to use that data.
- Close the **BufferedReader** object when you're done reading data from the file.
- Java Exceptions describes an alternative file input example that catches exceptions instead of using the *throws IOException* clause.
- If you do *catch* the exception, do not add the *throws IOException* clause to the method header.

6.3.6. String Tokenizer

When you are reading data into your program from a text file, you may need to interpret certain parts of each line differently. This is especially true if you need your program to create objects from text data contained in a file. Individual data items can be placed on different lines of the input data file, but this creates very long files. It is much more common to place all of the data items for one object on the same line and separate each item with some special character, called a *delimiter.* In this type of data file, each line of the file represents one record or one object. The example data file "student_scores.txt" shown below, contains data for three students and three exams scores for each student.

```
Kumar, Jatin: 87: 82: 95
Sharma, Ravi: 82: 89: 92
Verma, Pradeep: 93: 96: 100
```

Each line of input is interpreted as a string of characters for the name, that is followed by three integers for each object created. The delimiter character is the ':' in the above data file.

The process of splitting a line of text into different parts is known as *tokenizing,* and each piece is a *token.* The standard Java library includes a class called, **StringTokenizer** that makes this process much more convenient for Java programmers. Without this class, you would need to use nested repetition and selection statements to process each character one at a time and determine if it is part of the name or one of the exam scores and then store it accordingly. Java programmers simply need to learn how to construct a **StringTokenizer** object and what methods are available.

A **StringTokenizer** object is created from a **String** object. Use the ***hasMoreTokens*** method to determine if there are any remaining tokens to process. Use the ***nextToken*** method to return the next token as a **String** object. There is also a ***countTokens*** method that returns the number of remaining tokens available for a particular instance of the **StringTokenizer** class.

The **StringTokenizer** class is in the java.util package, so you must fully qualify the name as **java.util.StringTokenizer** or import the class. Here is a code fragment that reads the file shown above and computes the average score for each student listed. If you call ***nextToken*** when there are no tokens remaining, a **java.util.NoSuchElementException** will occur.

```
// Make sure that the file student_scores.txt exists and has
// valid data records. Otherwise, exceptions will occur.
File gradeFile = new File( "student_scores.txt" );
if ( gradeFile.exists() )
{
// Create the buffered reader for reading the file
BufferedReader inFile = new Buffered Reader(new FileReader(gradeFile ));
// Get the first line of the file
String line = inFile.readLine();

// If line is not end of file continue
while ( line != null )
{
// Create a StringTokenizer with a colon sign as a delimiter
java.util.StringTokenizer st = new java.util.StringTokenizer(line,":");

// Display the content of the first token
System.out.print( " Name: " + st.nextToken() );

// Display the total number of tokens remaining this string
int numScores = st.countTokens();

int sum = 0;
for ( int i=1; i <= numScores; i++ )
{
int score = Integer.parseInt( st.nextToken() );
sum += score;
}
System.out.println( " average = " + sum/numScores );

// Read next line of the file
line = inFile.readLine();

}

inFile.close();

}
```

6.3.7. File Output

Writing data to a file is similar to writing data to the screen. You will open a file for writing and then *print* to that file any data that you would like to store there. You must remember to *close the file* or risk having some data not be written and saved to the file. We will use each of these classes.

- java.io.FileWriter - used to open a file and connect an output stream to it.
- java.io.PrintWriter - used to write strings of characters (instead of bytes) to any Writer object.

When you intend to write data to a file, you should consider what the appropriate action to take is, if the file already exists. The safest option is to ask the user what to do, and then allow the user to choose overwrite the file, choose a different filename or cancel the operation. The example shown below assumes that the file opened by the **FileWriter** object will be overwritten if it already exists. If you do not want to overwrite the file if it already exists, then you must create and test a **File** object first. The ***exists*** method of the **File** class will return true if the file already exists.

```
// Create a FileWriter attached to a file named "out.txt".
// The second parameter sets whether or not new data
// will be appended to the end of the file or the beginning.
// false means that this file will be overwritten.
java.io.FileWriter fw = new java.io.FileWriter( "out.txt", false );
// Create a PrintWriter that automatically flushes data
// to the output file whenever the println method is used.
java.io.PrintWriter pw = new java.io.PrintWriter( fw, true );
// Buffer some data to write to the file (doesn't actually write until flush)
pw.print( "Some test data that will be written when flush is called.");
// Flush all buffered data to the file.
pw.flush();
// Write some data and automatically flush it to the file.
pw.println( data );
// Close the PrintWriter for added safety.
pw.close();
```

The nice thing about using a **PrintWriter** object is that you are already familiar with the ***print*** and ***println*** methods that are defined for all **PrintWriter** objects. By choosing the constructor that accepts a **String** for the filename and a boolean value, we are able to set the **PrintWriter** object so that it will always flush the data buffer on all calls to ***println***. This is safer, but less efficient. It is more efficient to wait until the buffer is full before writing to disk. In that case, the ***flush*** method must be called or the file must be closed to flush the remaining data from the buffer. (see more about files in chapter 14).

SUMMARY

In this chapter we have explained some additional features of java programming such as exceptional handling, Multithreading and Input/Output.

— Exceptional handling is a feature for handling the errors and exception at runtime using try and catch block.

— Multitasking is the ability of single processor to perform more than one operation at a time.

— Multhreading is the concept of multitasking.

— We have also explain the Input/Output console as well as file Input/Output as introduction.

SOLVED PROBLEMS

Q.6.1. What is user defined exception?

Ans. There are many exception defined by java which are used to track the run time exceptions and act accordingly. User can also define his exceptions that can be thrown in the same way java exceptions.

Q.6.2. What is the difference between Process and Threads?

Ans. Process is a program in execution. The execution of processes must be sequential. On the other hand thread is a light-weighted process which is executed in parallel with other threads. The theard shares some of the resources with other threads but ceratin things are unique to as particular thread like process id and program counter.

Q.6.3. What are three ways in which a thread can enter the waiting state?

Ans. A thread can enter the waiting state by invoking its sleep() method, by blocking on I/O, by unsuccessfully attempting to acquire an object's lock, or by invoking an object's wait() method. It can also enter the waiting state by invoking its (deprecated) suspend() method.

Q.6.4. What state does a thread enter when it terminates its processing?

Ans. When a thread terminates its processing, it enters the dead state.

Q.6.5. What's the difference between Thread and Runnable types?

Ans. A Java Thread controls the main path of execution in an application. When you invoke the Java Virtual Machine with the java command, it creates an implicit thread in which to execute the main method. The Thread class provides a mechanism for the first thread to start-up other threads to run in parallel with it.

The Runnable interface defines a type of class that can be run by a thread. The only method it requires is run(), which makes the interface very easy to fulfil by extending existing classes. A runnable class may have custom constructors and any number of other methods for configuration and manipulation.

Q.6.6. How does the run() method in Runnable work?

Ans. It may help to think of the run method like the main method in standard single threaded applications. The run method is a standard entry point to run or execute a class. The run method is normally only executed in the context of an independent Thread, but is a normal method in all other respects.

Q.6.7. What's the difference between a thread created as a subclass of java.lang.Thread and one created as an implementation of Runnable?

Ans. As threads, both behave the same. However, a thread that implements runnable can also extend another class.

Q.6.8. Why should synchronized blocks of code call wait()?

Ans. To give other waiting threads time to execute and prevent deadlock. Thread.yield() also works for this purpose

Q.6.9. Can there be a try block without a matching catch block?

Ans. Yes, if there is a finally block. try without catch or finally is a compile-time error.

Q.6.10. Can there be a catch block without a matching try block?

Ans. No. This is a compile-time error.

Q.6.11. Can there be a finally statement that is not attached to a try-catch block?

Ans. No. This is a compile-time error.

Q.6.12. When are exception objects created?

Ans. In general when an unexpected condition occurs. In specific, when you use throw to send a new exception, when the Java library uses throw to send a new exception, or when the Java runtime signals a RuntimeException.

Q.6.13. Does an exception have to be handled by the same try-catch block that threw it?

Ans. Not necessarily. If an exception isn't caught by the nearest try-catch block, it can be caught higher up. If it's a runtime exception it may not be caught at all.

Q.6.14. How can you make sure you catch every exception that a statement may throw?

Ans. Use this catch block as the last one in your hierarchy:

```
catch (Exception e) {

}
```

Q.6.15. If an exception isn't caught in a user-defined class, where does it go and what does it do?

Ans. If you're in an applet it may be caught in the applet viewer or web browser. However, the most likely result is that the thread in which the exception was thrown will be killed.

Q.6.16. Suppose you've defined an exception called NotANumberException(). Now suppose that NotAPositiveNumberException extends NotANumberException. Will

```
catch (NotANumberException) {
System.out.println("Caught a NotANumberException");
}
```

catch a NotAPositiveNumberException?

Ans. Yes. Superclasses always catch their subclasses.

Now suppose you have the following catch clause and throw a NotAPositiveNumberException. What happens?

```
catch (NotANumberException) {
    System.out.println("Caught a NotANumberException");
}
catch (NotAPositiveNumberException) {
    System.out.println("Caught a NotAPositiveNumberException");
}
```

"Caught a NotANumberException" is printed on System.out.

Q.6.17. In general which catch statement should come first? The one that catches the subclass or the one that catches the superclass? Why?

Ans. The one that catches the subclass. If the catch block that catches a superclass comes first, then the subclass will never be seen. You should move from the most specific (the subclass) to the most general (the superclass).

Q.6.18. What do you have to do to have an exception handled by both its own catch statement and by its superclass's catch statement?

Ans. Nest try statements and rethrow the exception inside the first catch block, like this

```
try {
    try {

        // statements that may fail
    }
    catch (SubclassException e) {
    // handle subclass
```

```
        throw(e);
        }
    }
    catch (superclass Exception e) {
    // handle superclass

    }
```

Q.6.19. What's the difference between throw and throws?

Ans. The throw keyword throws a new exception. The throws keyword declares that a method may throw a particular exception.

Q.6.20. Write a class that keeps a running total of all characters passed to it (one at a time) and throws an exception if it is passed a non-alphabetic character.

Ans.

```
    public class NotALetterException extends Exception {

        static int count=0;
        String msg;

        public NotALetterException(String s) {
        msg = s;
    }

    public static void countLetter(char c) throws NotALetterException {
    if (Character.isJavaLetter(c)) {
    count++;
    }
    else {
        throw new NotALetterException(c + "is not a letter");
    }
    }
    }
```

EXERCISES

Q.6.1. What is meant of exceptions?

Q.6.2. Explain use of try and catch with examples.

Q.6.3. What do you mean by user defined exceptions?

Q.6.4. Explain the Java defined exceptions?

Q.6.5. Differentiate throw and throws clause.

Q.6.6. What is the role of finally? When it is beneficial?

Q.6.7. Give an example of multiple catch clause.

Q.6.8. What do you mean by multithreading? Compare with multitasking.

Q.6.9. What is the difference between Thread and process?

Q.6.10. What do you mean by the life cycle of Thread?

Q.6.11. How the processor gives the priorities to the Threads?

Q.6.12. What is the role of is Alive () method?

Q.6.13. Give the details about the console output and input.

Q.6.14. Why we needed Stream objects?

Q.6.15. Explain the concept of string Tokenizer.

Chapter 7

I/O PROGRAMMING AND NETWORKING

CHAPTER OUTLINE

- File Handling
  - File Input
  - File Output
  - The File Class
- Networking
  - Internet Basic
  - IP-Datagram
  - Ports
  - Java Networking
  - URL Class
  - Reading file form URL
  - InetAddress Class
  - Sockets
  - Client/Server

7.1. FILE HANDLING

When inputting and outputting to the screen we make use of the System.in and System.out objects as discussed in chapter 12, System.in and System.out are special *stream* objects which read data from the keyboard and write data to the screen. Just see example given below:

Example 7.1: HELLO WORLD PROGRAM

```
import java.io.*;
class HelloWorld2
{
    // Create BufferedReader class instance
    static InputStreamReader input = new InputStreamReader(System.in);
    static BufferedReader keyboardInput = new BufferedReader(input);

    /* Main method */
    public static void main(String[] args) throws IOException
    {
    String name;

    System.out.print("What is your name? ");

    name = keyboardInput.readLine();
```

```
        System.out.print("\nHello " + name );
        System.out.println(" - Congratulations on writing your first" +
            " Java program which features some input!\n\n");
        }
    }
```

If we wish to read and write to a file we need to replace the in and out objects with appropriate alternative stream objects.

7.1.1. File Input

Before we can read from or write to a file we must first *open* the file. To open a file for reading we use the constructor contained FileReader class:

```
FileReader file = new FileReader("HelloWorld2.java");
```

This creates a stream object, called file, to read from a file called HelloWorld2.java (Example 7.1) contained in the current directory. If the named file cannot be opened a FileNotFoundException is generated. Consequently, whereas previously we declared instances of the InputStreamReader class as class instances we must now declare them as local instances within a method. Having done this we can now use the file object to create an instance of the BufferedReader class as before:

```
FileReader file = new FileReader("HelloWorld2.java");
BufferedReader fileInput = new BufferedReader(file);
```

Note the similarity between this and the declarations used in the code given in Example 2:

```
InputStreamReader input = new InputStreamReader(System.in);
BufferedReader keyboardInput = new BufferedReader(input);
```

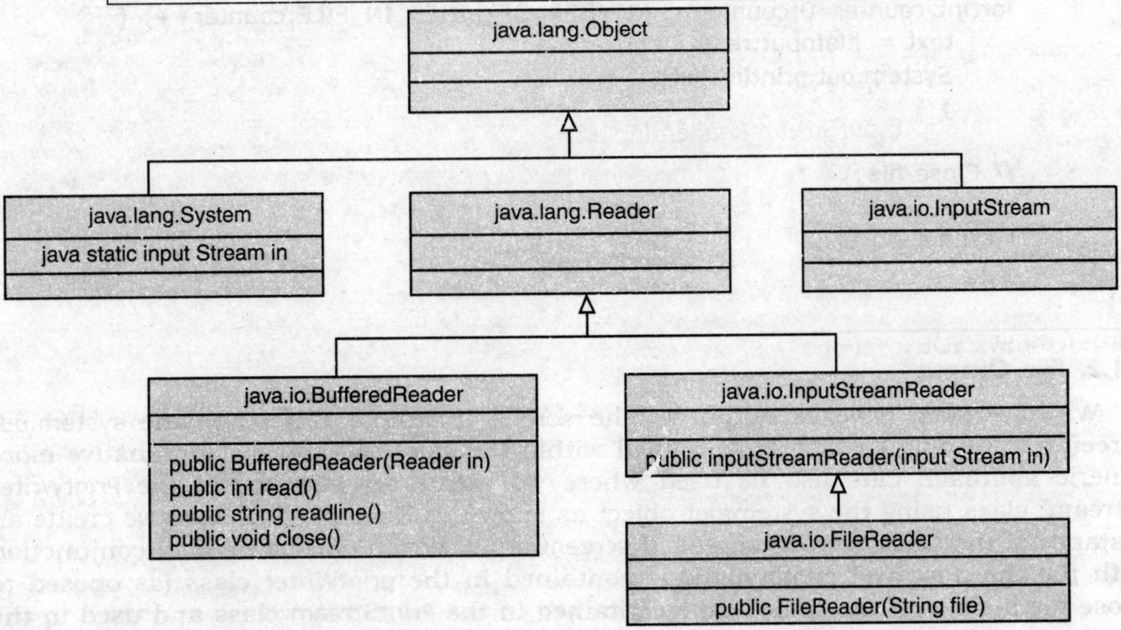

Fig. 7.1. Class diagram showing the connection between the different output classes

The BufferedReader constructor requires its argument to be an instance of the class Reader (or a sub-class of the class Reader), both the InputStreamReader and the FileReader classes are sub-callses of Reader and will therefore suffice. A class diagram illustrating the connections between these different classes is presented in figure 7.1.

Once we have finished reading from a file we should close it; this is achieved using the close instance method contained in the BufferedReader class. Because this is an instance method we must, of course, link it to an instance of the class BufferedReader:

```
fileInput.close();
```

The code presented in Example 7.2 reads the contents of the file HelloWorld2.java (presented in Example 7.1) and outputs the contents to the screen. Note that, in the Example 7.2, we keep reading from the file until we have read 31 lines.

Example 7.2: File Input Example

```
import java.io.*;

class FileIOExample1
{
    /* Main method */

    public static void main(String[] args) throws IOException
    {
        FileReader file = new FileReader("HelloWorld2.java");
        BufferedReader fileInput = new BufferedReader(file);
        String text;
        final int NUMBER_OF_LINES_IN_FILE = 31;

        // Read file and output
        for(int counter=0;counter < NUMBER_OF_LINES_IN_FILE;counter++) {
            text = fileInput.readLine();
            System.out.println(text);
            }

        // Close file

        fileInput.close();
        }
}
```

7.1.2. File Output

We have seen how we output to the screen (Example 7.1) using the system.out screen output object which is contained within the class System. An alteranative more generic approach can also be used where we create an instance of the PrintWriter "stream" class using the system.out object as shown in Example 7.3. Here we create an instance of the class PrintWriter, called screenOutput, which can be used in conjunction with the the print and printlnmethods contained in the printWriter class (as oposed to those methods with the same name contained in the PrintStream class and used in the example code in Example 7.1). The inter-relationship between the various output classes is illustrated by the class diagram presented in figure 7.2.

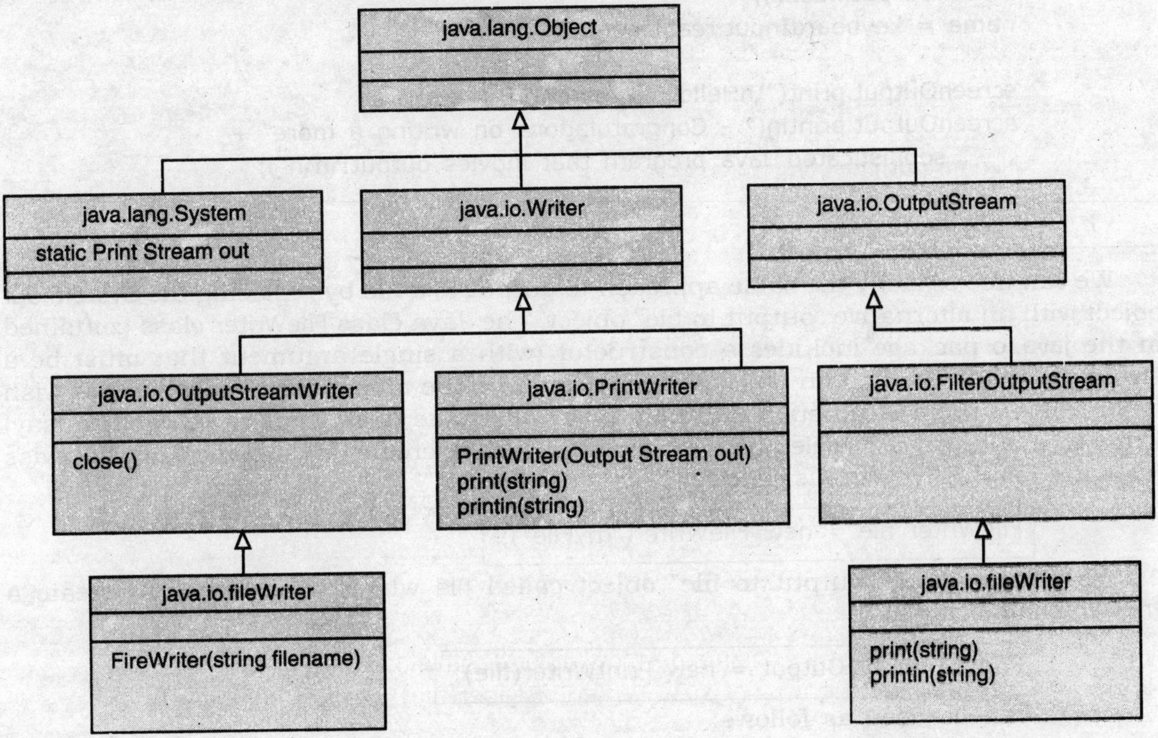

Fig. 7.2. Class diagram showing the connection between the different output classes

Example 7.3: Standard input/output

```
import java.io.*;

class HelloWorld4
{

    // Create BufferedReader class instance

    public static InputStreamReader input= new InputStreamReader (System.in);
    public static BufferedReader keyboardInput = new BufferedReader(input);

    // Crerste PrintWriter class instance

    static PrintWriter screenOutput = new PrintWriter(System.out,true);

    /* Main method */

    public static void main(String[] args) throws IOException
    {
```

```
        String name;
        screenOutput.print("What is your name? ");
        screenOutput.flush();
        name = keyboardInput.readLine();

        screenOutput.print("\nHello " + name );
        screenOutput.println(" - Congratulations on writing a more" +
            " sophisticated Java program that inovles output!\n\n");
    }
}
```

We can use esxactly the same approach to output to a file by replacing the System.out object with an alternative "output to file" object. The Java class FileWriter class contained in the java.io package includes a constructor (with a single argument that must be a string) that creates such an object. The argument is the name of the file which we wish to write to (if there is no such file then Java will create one). Thus if we wish to (say) write to a file called myFile then we would have to create an instance of the class FileWriter (lets call it file) as follows:

```
FileWriter file = new FileWriter("myFile");
```

We now have an "output to file" object called file which we can use to create a PrintWriter object:

```
PrintWriter fileOutput = new PrintWriter(file);
```

which can be used as follows:

```
fileOutput.print("Halo World")
```

or:

```
fileOutput.println("Halo World Again")
```

Thus to write to a file we replace the System.out stream object with an appropriate "file output" stream object. We create such an object using the FileWriter class constructor as shown above. However, the class FileWriter does not contain the instance methods print and println which we would like to use. These are contained in the PrintWriter class, thus we must also create an instance of this class.

We can now adapt the code given in Example 7.2 so that the contents of the file HelloWorld2 are output to a file called testFile; this is done in Example 7.4.

Example 7.4: File input and output example

```
import java.io.*;

class FileIOExample2
{

    /* Main method */

    public static void main(String[] args) throws IOException
```

```
    {
        FileReader file1 = new FileReader("HelloWorld2.java");
        BufferedReader fileInput = new BufferedReader(file1);
        FileWriter file2 = new FileWriter("testFile");
        PrintWriter fileOutput = new PrintWriter(file2);
        String text;
        final int NUMBER_OF_LINES_IN_FILE = 42;

    // Read file and output

    for(int counter=0;counter < NUMBER_OF_LINES_IN_FILE;counter++) {
        text = fileInput.readLine();
        System.out.println(text);
        fileOutput.println(text);
        }

        // Close file

        fileInput.close();
        fileOutput.close();
    }
}
```

7.1.3. The File Class

We express the location of a file using a *path name* (absolute or relative). The File class contains (amongst other things) a number of methods that can be used to check whether a named file exists or not. In Example 7.5, the code presented in Example 7.2 is repeated but with the inclusion of a method to check a file name, this time presented as a command line argument, using some of the methods contained in the File class.

Example 7.5: File input example with "file check"

```
import java.io.*;

class FileIOExample3
{

    /* Main method */

    public static void main(String[] args) throws IOException
    {
        String fileName = new String(args[0]);

        // CheckFile

        if (!checkFile(fileName)) System.exit(1);

        // Process
```

```
        FileReader file = new FileReader(fileName);
        BufferedReader fileInput = new BufferedReader(file);
        String text;
        final int NUMBER_OF_LINES_IN_FILE = 31;
        // Read file and output
        for(int counter=0;counter < NUMBER_OF_LINES_IN_FILE;counter++) {
            text = fileInput.readLine();
            System.out.println(text);
            }
        // Close file
        fileInput.close();
        }
    /* Check File */
    private static boolean checkFile(String fileName) {
        File src = new File(fileName);
        if (src.exists()) {
        if (src.canRead()) {
            if (src.isFile()) return(true);
            else System.out.println("ERROR 3: File is a directory");
            }
        else System.out.println("ERROR 2: Access denied");
        }
    else System.out.println("ERROR 1: No such file");
    return(false);
    }
}
```

7.2. NETWORKING

Networking architecture is based on the concept of layers. From Java Network Programming comes this layer model:

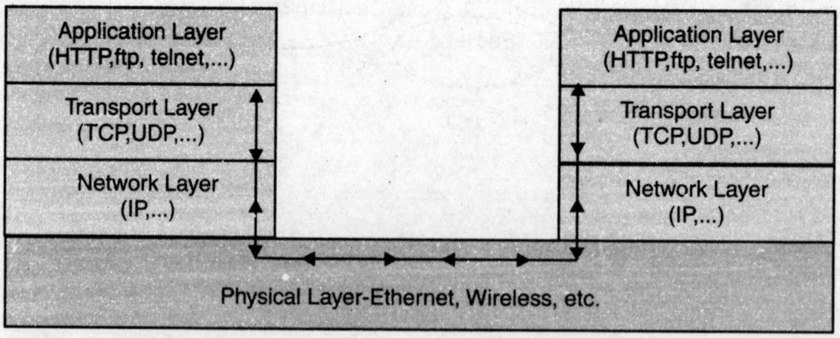

Fig. 7.3. Layer architecture

(A more formal OSI-Open System Interconnection - model has 7 layers but this one shows the essential layer definitions.)

Each layer has its own standardized protocols and applications programming interface (API), which refers to the functions, and their arguments and return values, called by the next higher layer.

Internally, the layers can be implemented in different ways as long as externally they obey the standard API.

For example, the Network Layer does not know if the Physical Layer is Ethernet or a wireless system because the device drivers respond to the function calls the same way.

The *Internet* refers primarily to the Network Layer that implements the Internet Protocol (IP) and the Transport Layer that implements the Transmission Control Protocol (TCP). In fact, you often here people refer to the "TCP/IP" network rather than calling it the Internet.

The application layer also includes various protocols, such as FTP (File Transport Protocol) and HTTP (Hypertext Transfer Protocol) for the Web, that rely on the TCP/IP layers.

Most users never look below the application layer. Most application programmers never work below the TCP/IP layers.

7.2.1. Internet Basic

As mentioned in the above, the Internet refers to the TCP & IP protocols used in the Transport and Network layers, respectively.

The diagram below shows how the TCP and IP layers split an email into groups of bytes called *packets*. A packet contains a *header* with the destination and source addresses and miscellaneous other information such as error correction bytes. (We will discuss the IP address format later.) The body of the packet consists of the *data payload*.

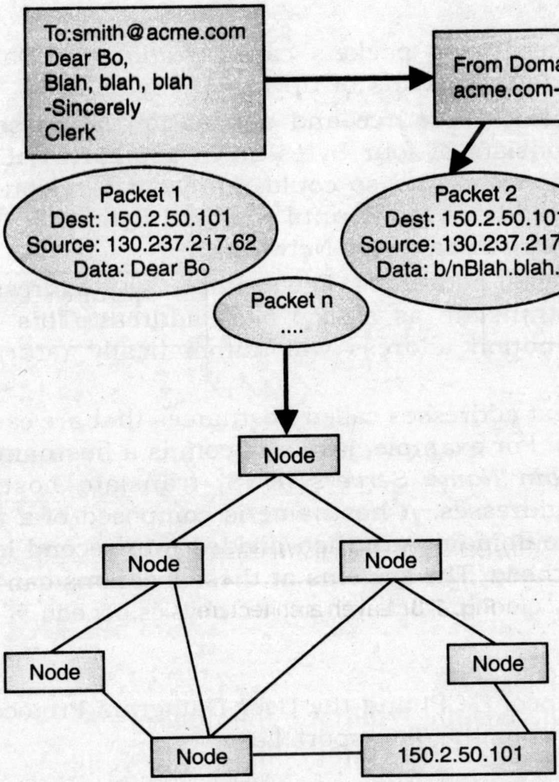

Fig. 7.4.

These packets are sent through the network towards there destination. At the *nodes* of the network, there are devices called routers. The routers read the destination addresses on the packets and send the packet to the next node closest to the final destination, just as a letter in the postal delivery system is passed along to the next node nearest the final address. If a node doesn't respond or if the traffic load needs balancing, the transmitting node looks for alternate routes.

When the packets reach their final destination, the original message is rebuilt. The packets may arrive out of order so the rebuilding must wait for all packets to arrive.

The above figure sketches the packet's travels through the network. A node in the local area network (LAN) could be another computer or a dedicated *router*, which is essentially a computer dedicated to packet transmission. A router receives packets from different nodes and sends them to the proper destination. Typically the local router sends the packet to an internet service provider (ISP), such as a DSL or cable internet system, which is connected to the *Internet backbone*.

The backbone uses very high speed, high capacity routers and fibers to send the packet to the ISP closest to the destination. From there the packet works its way to the destination computer, which might lie inside a LAN, which might in turn connect to a WAN (wide area network) such as on a computer campus.

The header consists of a destination address, source address, and other miscellaneous information such as error correction bytes. Routers read the destination address and send the packet to the next link determined to be the shortest path.

This path can vary if, for example, the router determines that a previously chosen link has too much traffic or the router at the other end of the link is not responding.

7.2.2. IP-Datagram

The IP layer communicates using packets called *datagrams*. Datagrams contain headers of 20-60 bytes and data payloads of up to 65K bytes.

Datagrams contain the basic IP source and destination addresses, where the IP (Internet Protocol) address consists of four bytes as in 130.237.217.62. The left most byte is the highest order addressing and so could represent a region or country. The lower order bytes narrow the address down until the final byte indicates, for example, a particular computer on a LAN (Local Area Network).

There are a number of special addresses. For example, any address beginning with 127 such as 127.0.0.1 will translate as a *loop back* address. This means that any packets sent with this destination address will automatically return to the source computer.

Users typically deal with text addresses called hostnames that are easier to remember than the numerical addresses. For example, java.sun.com is a hostname. Special nodes on the Internet called *Domain Name Servers* (DNS) translate hostnames into the corresponding numerical IP addresses. A hostname is composed of a top level domain such as com, se, or edu. These domains are then divided into second level subdomains such as sun.com , kth.se, or ucr.edu. The systems at these locations can divide a domain further such as java.sun.com, gluon.particle.kth.se, or physics.ucr.edu.

7.2.3. TCP-UDP

Transmission Control Protocol (TCP) and the User Datagram Protocol (UDP) are two separate Internet protocols within the Transport Layer.

Datagrams from the same message can travel completely different paths as the routers dynamically choose paths for the same destination address so as to avoid loading down any one link. Thus datagrams may become lost or arrive out of order from how they were sent.

The Transport Layer attempts to smooth over the problems of the IP layer. The Transport Layer can place packets into the proper order and request *retransmission* of missing packets (that is, packets not arriving within a given time).

TCP guarantees bytes in correct order. UDP does not guarantee correct order but has lower overhead than TCP.

For some applications, such as audio and video transmission, the loss of a few bytes is not noticeable and so the UDP serves very well since it is faster than TCP.

Application Layer

The Application Layer involves all the user related programs like web browsers and ftp. The application layer programs use various protocols:

- **HTTP** - Hyper Text Transfer Protocol
- **FTP** - File Transfer Protocol
- **Telnet** - the telnet protocol provides for console sessions.
- **SMTP** - Simple Mail Transport Protocol - email.

These protocols require TCP since data can not be randomly dropped from source files and web pages. Other application layer programs like ping, which sends test packets to a given IP address, can use the simpler UDP. Most Java network programs only deal with this application layer.

7.2.4. Ports

While the IP address will get the packet to the right computer, how does the packet get to the right program? A computer might have:

- Several programs running at the same time
- Several programs trying to communicate via the Internet
- All the programs communicating out the same physical Ethernet cable

The *Port* number is the key for organizing all of this. The packets are guided to the correct program by the operating system according to the port number as indicated in this diagram:.

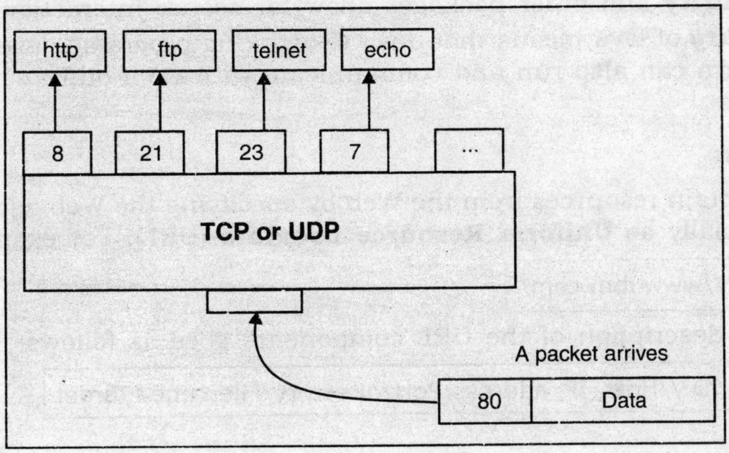

Fig. 7.5. Standard ports

The port is a 16-bit number. It is occasionally seen on Web addresses:

```
http://www.someplace.com:80/
```

where 80 is the default for the HTTP server. Nowadays the port number is seldom included in the URL since the default is almost always used.

Various applications using particular protocols use standard port values as shown in the above figure 7.5. Unix machines reserve ports 1-1023 for standard services. Windows machines do not restrict the use of these ports but to make your Java programs portable, it is wise to use port values above 1023.

One type of firewall assigns port numbers to the machines behind its shield. Incoming packets all go to the same IP address but with different port numbers according to the machine they are destined for. This both reduces the exposure and accessibility to the machines and reduces the need for universally unique IP numbers. Behind the firewall, the local addresses can be the same as in other LANs since the LANs are isolated from each other.

This is one of the reasons that four bytes remain sufficient for the internet despite the explosion in the number of devices with IP addresses. (Nevertheless, a sixteen byte version of IP addressing called IPv6 will gradually take over.)

7.2.5. Java Networking

Java was designed from the start with networking in mind. Java became famous because of? applets, which were invented to provide dynamic content to web pages. Many other features of Java, however, extend Java network programming far beyond mere applets. Some of these include:

- **java.net** package in the core language includes numerous network related classes.
- ***Remote Method Invocation* (RMI)** packages allow a Java program running on one machine to invoke methods on another machine.
- ***Streaming I/O*** architecture provides a uniform structure in that I/O over a network is treated the same as I/O to a disk or the console.
- ***Serialization*** allows objects to be sent over a network and reconstructed on another machine.
- ***Threading*** allows easier creation of server programs where threads can be easily spun off to serve new clients as they connect.
- **java.security** and other packages allow for secure interactions over networks.
- ***Portability*** of java means that Java networking programs developed on one type of platform can also run and communicate with each other when used on other platforms.

7.2.6. URL Class

Browsers obtain resources from the Web by specifying the Web addresses, which are referred to officially as **Uniform Resource Locators** (URL). For example,

```
http://www.ibm.com/
```

The formal description of the URL components goes as follows:

```
Protocol_ID://Host_IP_address:Port/directory/Filename#Target
```

where

Protocol_ID = HTTP, FTP, etc

Host_IP_address = Host name in either numerical IP or hostname format

Port = default is 80

Filename = Name of a hypertext web page or other type of file. (Default index.htm[l])

Target = optional reference address within a Web page

In Java we can create an instance of the java.net.URL class. These specify the location of a file on the network. We have, for example, used URL objects to obtain the location of image files in applets as shown here:

```
getImage( getCodeBase(), filename);
```

The applet method getCodeBase() returns a URL object for the location of the applet file. Here it is assumed the image file resides in the same directory

The URL class provides several methods for obtaining URL information ,For Example:

Example 7.8: Parse a URL address

```
import java.net.*;
import java.io.*;
/** Parse a URL address into its components. **/
public class ParseAddress
{
    public static void main (String[] args) {
    if (arg.length !=1) {
        System.out.println ("Error: missing the url argument");
        System.out.println ("Usage: java ParseURL <url>");
        System.exit (0);
    }

    try {
        URL url = new URL (arg[0]);
        System.out.println ("Protocol = " + url.getProtocol ());
        System.out.println ("Host = " + url.getHost ());
        System.out.println ("File name = " + url.getFile ());
        System.out.println ("Port = " + url.getPort ());
        System.out.println ("Target = " + url.getRef ());
    }
    catch (MalformedURLException e) {
    System.out.println ("Bad URL = " + arg[0]);
    }

}
}
```

You run the program as follows:

```
c:\> java ParseAddress http://www.myschool.edu:80/data/x.html#d
```

This program outputs:

```
Protocol = http
Host = www.myschool.edu
File name = /data/x.html
```

Port = 80
Target = d

Also, URL's can be constructed from a URL string

```
URL kth = new URL ("http//www.kth.se/index.html");
```

or from separate protocol, host, file strings:

```
URL kth = new URL ("http","www.kth.se","/index.html");
```

When you attempt to create a URL object, the constructor will validate the components . If they are in error, it will throw a MalformedURLException. So you must catch this exception from the URL constructors.

7.2.7. Reading file from URL

Often one of the first applets that new Java programmers want to write for their own work involves reading a file of some sort. This could be simple database information such as a list or table of values and labels in text format.

We have learned to read image files and audio files from an applet but have not yet read a text file. This is actually fairly easy to do. However, keep in mind that for applets the SecurityManager restricts access to only those files accessible from the location of the web page on which the applet code resides.

Below we show the applet ReadFileViaURL that reads a text file. Note that we use the URL class that we discussed in the previous section. In this case, the URL class provides a method to obtain file access via an InputStream object.

```
// Open a stream to the file using the URL.
try {
    InputStream in=url.openStream ();
    BufferedReader dis =
    new BufferedReader (new InputStreamReader (in));
    fBuf = new StringBuffer () ;

    while ( (line = dis.readLine ()) != null) {
    fBuf.append (line + "\n");
    }
    in.close ();
}
catch (IOException e) {
    fTextArea.append ("IO Exception = "+e);
    System.out.println ("IO Exception = "+e);
    return;
}
```

We wrap this stream with an InputStreamReader to provide proper character handling. We in turn wrap this stream with an BufferedReader, which provides buffering to smooth out the stream flow and also contains the convenient readLine() method that grabs an entire line and returns it as a string.

The program can also run as a standalone application.

Example 7.9: Reading a Local File with a URL

```
import javax.swing.*;
import java.awt.*;
import java.awt.event.*;
import java.util.*;
import java.io.*;
import java.net.*;

public class ReadFileViaURL extends JApplet implements ActionListener
{
  // A Swing text area for display of string info
  JTextArea fTextArea = null;
  String fFileToRead ="data.txt";
  StringBuffer fBuf = null;

  // Flag for whether the applet is in a browser
  // or running via the main () below.
  boolean fInBrowser = true;

  /** Build the GUI
  * Create a User Interface with a text area with scroll bars
  * and a Go button to initiate processing and a Clear button
  * to clear the text area.
  **/
  public void init () {
  Container content_pane = getContentPane ();

  JPanel panel = new JPanel (new BorderLayout ());

  // Create a text area to display file contents
  fTextArea = new JTextArea ();
  fTextArea.setEditable (false);

  // Add to a scroll pane so that a long list of
  // computations can be seen.
  JScrollPane area_scroll_pane = new JScrollPane (fTextArea);

  panel.add (area_scroll_pane,"Center");

  JButton go_button = new JButton ("Go");
  go_button.addActionListener (this);

  JButton clear_button = new JButton ("Clear");
  clear_button.addActionListener (this);
  JButton exit_button = new JButton ("Exit");
  exit_button.addActionListener (this);
  JPanel control_panel = new JPanel ();
```

```
    control_panel.add (go_button);
    control_panel.add (clear_button);
    control_panel.add (exit_button);

    panel.add (control_panel,"South");

    // Add text area with scrolling to the contentPane.
    content_pane.add (panel);

    // If running in a browser, read file name from
    // applet tag param value. Else use the default.
    if (fInBrowser) {
    // Get setup parameters from applet html
    String param = getParameter ("FileToRead");
    if (param != null) {
    fFileToRead = new String (param);
    }
  }
  } // init

    /** Use a URL object to read the file **/
    public void readFile () {
    String line;
    URL url=null;

    // Get the URL for the file.
    try {
        if (fInBrowser)
            url = new URL (getCodeBase (), fFileToRead );
    else {
    File file = new File (fFileToRead);
    if (file.exists () )
    url = file.toURL ();
    else {
        fTextArea.append ("No file found");
        System.out.println ("No file found");
        System.exit (0);
        }
    }
  }
  catch (MalformedURLException e) {
    fTextArea.append ("Malformed URL = "+e);
    System.out.println ("Malformed URL= "+e);
    return;
  }

    // Open a stream to the file using the URL.
    try {
```

```
      InputStream in=url.openStream ();
      BufferedReader dis = new BufferedReader (new InputStreamReader
      (in));
      fBuf = new StringBuffer () ;

      while ( (line = dis.readLine ()) != null) {
      fBuf.append (line + "\n");
    }

    in.close ();
  }
    catch (IOException e) {
    fTextArea.append ("IO Exception = "+e);
    System.out.println ("IO Exception = "+e);
    return;
  }

    // Load the file into the TextArea.
    fTextArea.append (fBuf.toString ());
    } // readFile

    /**
    * Can use the start () method, which is called after
    * init () and the display has been created.
    **/
    public void start () {
    // Now read the file.
    readFile ();
    }

    /** Respond to the buttons **/
    public void actionPerformed (ActionEvent e) {

    String source = e.getActionCommand ();
    if (source.equals ("Go"))
    start ();
    else if ( source.equals ("Clear") )
    fTextArea.setText (null);
    else
    System.exit (0);

    } // actionPerformed

    /** Display the println string on the text area **/
    public void println (String str) {
    fTextArea.append (str + '\n');
    }
```

```
/** Display the print string on the text area **/
public void print (String str){
fTextArea.append (str);
}
/** Create the frame and add the applet to it **/
public static void main (String[] args) {
int frame_width=200;
int frame_height=300;
// Create an instance of ReadFile_JApplet8 and add to the frame.
ReadFileViaURL applet = new ReadFileViaURL ();
applet.fInBrowser = false;
applet.init ();
// Following anonymous class used to close window & exit program
JFrame f = new JFrame ("Read file from URL Demo");
f.setDefaultCloseOperation (JFrame.EXIT_ON_CLOSE);

// Add applet to the frame
f.getContentPane ().add ( applet);
f.setSize (new Dimension (frame_width,frame_height));
f.setVisible (true);
} // main

} // class ReadFileViaURL

/* With following applet tags, you can use this source with
appletviewer:
    <applet code=<applet code="ReadFileViaURL.class" width=350
    height=350>
    </applet>
*/
```

The program can also run as a standalone application. We see in the readFile() method of ReadFile_JApplet8 how to obtain a URL object from a File object for local files.

7.2.8. InetAddress Class

The java.net.InetAddress class represents IP addresses. It works with either the hostname (e.g. www.kth.se) or the numerical IP address (e.g. 64.124.81.56) of an Internet host. InetAddress class offers a number of useful methods for dealing with hostnames and IP addresses.

The Example 7.10 program below illustrate some of the capabilities of the InetAddress class.

The applet LocalAddress displays the local host and IP address. Note that the SecurityManager in the browser JVM may block access to this information. You can try it in the appletviewer tool or run it as an application.

Example 7.10: Show how InetAddress can provide the local IP address

```
import java.applet.*;
import java.awt.*;
```

```
import java.net.*;
public class LocalAddress extends Applet
{
    String fMessage = "";
    /**
    * Create an instance of InetAddress for the
    * local host and display the local IP address.
    **/
public void init () {
    try {
    InetAddress local_Address = InetAddress.getLocalHost ();
    fMessage ="Local IP address = " + local_Address.toString ();
    }
    catch (UnknownHostException e) {
    fMessage ="Unable to obtain local IP address";
    }
    System.out.println (fMessage);
    } // init

    /** Paint IP info in Applet window. **/
    public void paint (Graphics g) {
    g.drawString (fMessage,20,20);
}
    /** Print out the local address in app mode.**/
    public static void main (String [] args) {
        LocalAddress applet =
        new LocalAddress ();
        applet.init ();
    } // main
} // Class LocalAddress
```

The following Example 7.11, returns the IP address if given a host name, or returns the host name if given the IP address.

Example 7.11: IP argument to a host name or a host name to an IP address.

```
import java.net.*;
class TranslateAddress {
    public static void main (String args[])
    {
    // Look for command line argument
    if (args.length != 1) {
    System.out.println ("Error! No IP or host name address");
    System.out.println ("Usage: java TranslateAddress java.sun.com");
    System.out.println ("    or java TranslateAddress 209.249.116.143");
    System.exit (0);
}
    try {
```

```
    // When the argument passed is a host name (e.g. sun.com),
    // the corresponding IP address is returned. If passed
    // an IP address, then only the IP address is returned.
    InetAddress address = InetAddress.getByName (args[0]);
    System.out.println ("Address " + args[0] + " = " + address);

    // To get the hostname when passed an IP address use
    // getHostName (), which will return the host name string.
    System.out.println ("Name of " + args[0] + " = " +
                address.getHostName ());
    } catch (UnknownHostException e) {
   System.out.println ("Unable to translate the address.");
  }

 } // main

}
```

Sample runs of this Example program:

```
c:\> java TranslateAddress gluon.particle.kth.com
Address gluon.particle.kth.se = gluon.particle.kth.se/130.237.34.133
Name of gluon.particle.kth.se = gluon.particle.kth.se

c:\> java TranslateAddress 130.237.34.133
Address 130.237.34.133 = /130.237.34.133
Name of 130.237.34.133 = gluon.particle.kth.se
```

Note that getByName(String str) returns an instance of InetAddress when given a host name. Then when this objects converts to a string it displays the HostName/IP_Address pair.

However, when given an IP address the resulting InetAddress object does not show the host name. Instead you must use the the getHostName() method to obtain the hostname.

7.2.9. Sockets

Sockets provide connections between applications and allow streams of data to flow. Sockets are fairly straightforward to set up in Java. Java provides two kinds of sockets:

1. **Socket:** provides a connection-oriented protocol that behaves like telnet or ftp. The connection remains active, even with no communications occurring, until explicitly broken.
2. **DatagramSocket:**
   - a connectionless protocol
   - transfers datagram packets
   - no fixed connection
   - does not keep packets in order
   - no guarantee a packet will arrive at its destination

The class ServerSocket is not a socket , but it monitors a given port and returns a Socket object when a client attempts to connect to that port.

Concept

Sockets provide connection between applications of clients/server and allow streams of data to flow.

The Example program 7.12 shown below, illustrates how to run the *whois* internet operation that returns information about a given domain name. The registry site whois.internic.net provides this service from anywhere on the internet.

The code segment shows how to create a socket that connects a stream to port 43 at the remote site. An output stream is obtained from the socket and a PrintWriter wrapped around it. This is then used to send a whois query to that socket.

Similarly, an input stream is obtained from the socket and wrapped first with an InputStreamReader and then a BufferedReader. The latter class provides the readLine() routine or obtained a whole line of text at one time from the whois output. ?You can enter other domain names in the text field.

Though written in applet form, the security manager in most browser JVMs will block accesses to IP addresses other than the source of the applet. So we only show an image of a typical output. You can, however, run the program as an application or an appletviewer.

Example 7.12: The whois output in a text area

```
import javax.swing.*;
import java.awt.*;
import java.awt.event.*;
import java.util.*;
import java.io.*;
import java.net.*;

public class WhoisApplet extends JApplet  implements ActionListener
{
   // A Swing textarea for display of string info
   JTextArea fTextArea = null;
   JTextField fTextField = null;
   String fFileToRead="data.txt";

   //Buttons
   JButton fGoButton;
   JButton fClearButton;
   JButton fExitButton;

   // Flag for whether the applet is in a browser
   // or running via the main () below.
   boolean fInBrowser = true;
   public void init () {

   // Create a User Interface with a textarea with sroll bars
   // and a Go button to initiate processing and a Clear button
   // to clear the textarea.
```

```
Container content_pane = getContentPane ();
JPanel panel = new JPanel (new BorderLayout ());

// Create an instance of DrawingPanel
fTextArea = new JTextArea ();
fTextArea.setEditable (false);
fTextArea.setLineWrap (true);

// Add to a scroll pane so that a long list of
// computations can be seen.
JScrollPane area_scroll_pane = new JScrollPane (fTextArea);

panel.add (area_scroll_pane,"Center");
fTextField = new JTextField ("sun.com", 10);
// If return hit after entering text, the
// actionPerformed will be invoked.
fTextField.addActionListener (this);

fGoButton = new JButton ("Go");
fGoButton.addActionListener (this);

fClearButton = new JButton ("Clear");
fClearButton.addActionListener (this);

fExitButton = new JButton ("Exit");
fExitButton.addActionListener (this);

JPanel control_panel = new JPanel ();

control_panel.add (fTextField);
control_panel.add (fGoButton);
control_panel.add (fClearButton);
control_panel.add (fExitButton);

panel.add (control_panel,"South");
// Add text area with scrolling to the contentPane.
content_pane.add (panel);

} // init
/** Connect to the whois service via a socket. Write the
* address with a PrintWriter object and read the output
* via a BufferedReader. Send the output to the text area.
**/
public void whoisConnect ( String queryAddress) {

int port = 43; // Standard whois port
String reply;
```

```
try {
Socket whois_socket = new Socket ("whois.internic.net", port);

PrintWriter print_writer =
new PrintWriter ( whois_socket.getOutputStream (),true);
print_writer.println (queryAddress );

InputStreamReader input_reader =
new InputStreamReader (whois_socket.getInputStream ());
BufferedReader buf_reader = new BufferedReader (input_reader);

while ( (reply = buf_reader.readLine ()) != null) {
// Write the whois info to the textarea.
fTextArea.append (reply + '\n');

}
fTextArea.append ("\n\n"); // Add space between queries
} catch (IOException e ){
fTextArea.append ("IO Exception = "+e);
System.out.println ("IO Exception = "+e);
return;
}
} // whoisConnect
/** Respond to the buttons **/
public void actionPerformed (ActionEvent e) {

Object source = e.getSource ();
if (source == fGoButton || source == fTextField ) {
String query = fTextField.getText ();
if (query == null || query.length () <=1 )
fTextArea.setText ("Error: Bad host entry");
else
    try {
    whoisConnect (query);
    }
catch (Exception ec) {
fTextArea.append("\n\n Error attempting to connect to WhoIs!\n");
fTextArea.append("Exception = \n" + ec);

} else if (source == fClearButton)
    fTextArea.setText (null);
else if (!fInBrowser)
    System.exit (0);
} // actionPerformed
/** Display the println string on the text area **/
public void println (String str) {
    fTextArea.append (str + '\n');
}
```

```
    /** Display the print string on the text area **/
    public void print (String str){
    fTextArea.append (str);
}
    public static void main (String[] args) {
    //
    int frame_width=450;
    int frame_height=300;

    //
    WhoisApplet applet = new WhoisApplet ();
    applet.fInBrowser = false;
    applet.init ();

    // Following anonymous class used to close window & exit program
    JFrame f = new JFrame ("whois Demo");
    f.setDefaultCloseOperation (JFrame.EXIT_ON_CLOSE);

    // Add applet to the frame
    f.getContentPane ().add ( applet);
    f.setSize (new Dimension (frame_width,frame_height));
    f.setVisible (true);
    } // main

}// class WhoisApplet
```

Note: WhoisApplet operations are blocked by the JVM security manager in most browsers but you can run it in standalone mode instead.

Below we show an image of an example of the program in action:

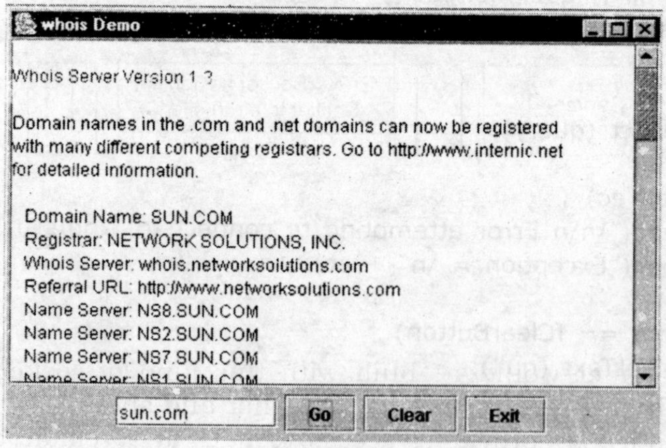

Fig. 7.6.

7.2.10. Client/Server

The **Client/Server** paradigm has become a dominant one for the Internet. In this model, the clients are programs running on remote machines that communicate with a program called the server that runs at a single site and responds to requests from many clients. The server provides the clients with, say, Web pages or database information.

Much of the World Wide Web is built on the client/server paradigm.The clients are Web browsers run by many millions of individual users, and the servers are the many Web hosting systems running at the many host sites on the Web.

A single server at a single host can support many hundreds or thousands or more of clients from around the world. Large systems that serve hundreds of thousands of clients balance the server load over multiple machines in an arrangement called "server farms."

With Java you can build client/server systems with sockets or with RMI (Remote Method Invocation).

In a socket based client/server system, a server listens to a particular port for client applications sending requests for connections. A ServerSocket class is provided in Java that allows for a server to monitor and answer such requests for connections. The client sends the request for a connection by creating a socket with the host name and port for that server as discussed in the previous section.

The figure 7.7 below shows a diagram illustrating the basics of a socket-based client/server system. The ServerSocket instance listens for a client to connect to the particular port. When a client request arrives, the ServerSocket object sets up a Socket instance for the connection and then spins off a new thread to interact with the client via that socket. Many clients can therefore be served since each client has an independent thread dedicated to it.

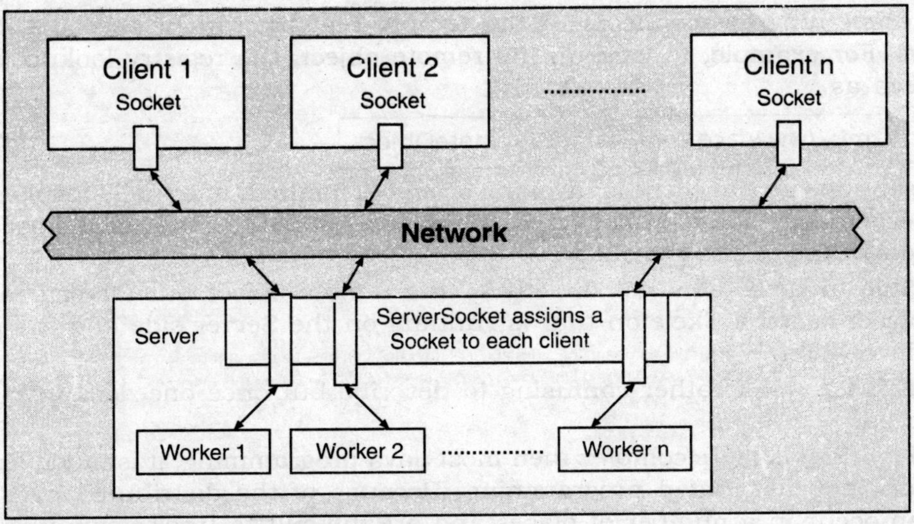

Fig. 7.7.

A Java Web server would be built with this kind of socket oriented client/server approach. We will discuss this approach further and show how to build custom servers for applications such as providing access to data at a remote device.

7.2.11. RMI

The RMI system in Java allows an object running in one JVM on one host to invoke the method of a remote object that is running in another VM (Virtual Machine) on another host just as if they were both in the same JVM on the same platform. To do this, RMI allows objects to be passed back and forth (when the objects are Serializable) over the network.

Sockets involve network communications at a fairly low level that require detailed handling of the socket setup, protocols, etc. RMI provides a higher level of networking where much of the networking machinery is hidden. An object calls methods of the remote object just as if the remote object was running locally.

> **Concept**
>
> *RMI provides a mechanism to invoke a method of an object that is being executed on a remote machine as a part of distributed computing.*

Typically, RMI setups are discussed in terms of a client that runs methods on a server on a remote host. (However, there's no reason the *server* could not invoke methods on the *client*, as well. So the terminology can be somewhat misleading.)

Setting up RMI programs requires several steps. On the server side the objects need to implement a sub-class of the Remote interface that implemented by the objects to be remotely invoked. Also, the objects must extend UncastRemoteObject.

The Registry program on the server side must be activiated. It is what the client interacts with to find the objects of interest. The server must register or bind the name of its object with the registry.

A special RMI compiler called **rmic** is run on the server to create the Stub and Skeleton classes that carry out the lower level interaction.

The client program connects to the remote registry with a special URL protocol called rmi. For example, to look up the remote object, the registry lookup is passed a string such as

```
rmi://www.ace.com/NameOfRemoteObject.
```

While it appears the client is invoking a remote method, it actually invokes a method on a special local object called a Stub that is passed to the local host when the connection to the Registry is made.

The Stub in turn talks not directly to the remote object but instead talks with a special object called a Skeleton that is running on the Server side and carries out the method invocation.

All of this can get rather confusing to describe but once one does it in practice it is not so bad.

Although RMI is more complex then most Java programming, it is much less complex then traditional distributed programming. Because of the distributed nature of RMI, errors can occur in a number of places and are difficult to track down until you have gained experience and insight. You will need at least three programs running at the same time, (1) the rmiregistry, (2) the program's server, and (3) a client. Each has their own potential for error. The code examples which follow have been tested and work on multiple platforms.

As with all programming endeavors, it is best to first understand the objective, then create a specification and plan of attack BEFORE writing any code. RMI can become quite complex so this approach is essential.

Project Description

We will create a distributed random quote client/server application. We need to supply a client which invokes methods on a remote object located on the server. The server will compute and return a response which will be displayed by the client.

Project Specification

The client will provide simple user interaction and display from one to ten unique random quotes. The server will manage the quote resources, compile random quotes, and return the result to the client. The client is responsible for displaying the content. We will use the default RMI port 1099 and let RMI handle our socket interaction behind the scenes.

The components that we will need to create are:

1. A **RandomQuote** interface
2. A **QuoteServer**
3. A **QuoteClient**
4. A **Stub** and a **Skeleton** (automatically generated)
5. Additionally we will need our Quote Resources.

Now that we have a clear picture of what we want, we can write the interface. This interface provides the "blueprint" which will be implemented by the server and used by the client.

```
import java.rmi.*;

public interface RandomQuote extends Remote {
    public String getQuote() throws RemoteException;
public String getQuotes(int n) throws RemoteException;

}
```

We must import java.rmi.* and implement java.rmi.Remote. Additionally, since we are using a distributed environment, we need to handle remote errors by declaring each method to throw a RemoteException.

Now we need to implement this interface and create the QuoteServerImpl. This is the class which will do the "heavy duty" processing. The QuoteServerImpl's function is to manage the quote resources (load them if necessary) and generate a formatted response of single or multiple random quotes. How it does this is irrelevant to the client which just needs to know what methods it can invoke and the data type of the return values. We don't want to produce duplicates if multiple quotes are requested so we design the QuoteServerImpl to make sure that this does not happen. The implementation then compiles the quotes and returns the result. Note that only the methods which are described in the interface are available to the client regardless of their visibility level.

```
import java.rmi.*;
import java.rmi.server.UnicastRemoteObject;
import java.io.*;
import java.util.*;
public class QuoteServerImpl implements RandomQuote {
    ArrayList quotes = new ArrayList(); //Java 2, use a Vector for 1.1.
    ArrayList chosen = new ArrayList();
```

```
    String resources = "SET THIS!!!!!"; //full path to your resources!!

  public QuoteServerImpl() throws RemoteException {
    UnicastRemoteObject.exportObject(this);
  }
 public String getQuote() throws RemoteException {
    if (quotes.size() == 0) {
    readQuotes();
  }
    String output = getRandomQuote();
    return output;
  }

  public String getQuotes(int n) throws RemoteException {
        if (quotes.size() == 0) {
        readQuotes();
    }
    chosen.clear();
    StringBuffer output = new StringBuffer();
    String temp = "";
    for (int i = 0; i < n; i++) {
    temp = getRandomQuote();
    if (chosen.contains(temp)) {
        i--;
    } else {
        output.append(temp + "\n\n");
    chosen.add(temp);
    }
  }
 return output.toString();
 }

 private void readQuotes() {
   quotes.clear();
   try {
       File f = new File(resources);
       FileReader fr = new FileReader(f);
       BufferedReader in = new BufferedReader(fr);
       String s = "";
       while ((s = in.readLine()) != null) {
            quotes.add(s);
       }
       in.close();
   } catch (IOException e) {
       System.out.println("ERROR: " + e);
       quotes.add("Unable to access the Quotes" +
       "at this time. Please try again later.");
```

```
    }
  }

  private String getRandomQuote() {
      int n = quotes.size();
      int selection = (int)Math.round(Math.random() * n);
      return (String)quotes.get(selection);
  }
}
```

The QuoteServerImpl implements our interface which means it extends Remote. Since we are already extending Remote we cannot also extend UnicastRemoteObject (single inheritance), so we use the convenient static method UnicastRemoteObject.exportObject(Remote obj) to export a reference to the server. Other than this, the QuoteServerImpl looks much like any other Java class.

Now we will write the actual server which binds to the **rmiregistry** and makes the remote object available.

```
import java.net.*;
import java.rmi.*;

public class QuoteServer {

    public static void main(String[] args) {
        try {
    QuoteServerImpl q = new QuoteServerImpl();
    Naming.rebind("quotes", q); // sets the handle
    System.out.println("QuoteServer started and "+ "awaiting connections.");
  }
  catch (RemoteException er) {
    System.out.println("Exception in QuoteServer.main: " + er);
  }
  catch (Exception err) {
      System.out.println( "Exception occurred: " + err);
    }
  }
}
```

This looks similar to a normal main, however, note that it binds the implementation to the handle, "quotes", which is used by all clients that need to find the remote object in the **rmiregistry**.

Now we need to write the simple client.

```
import java.rmi.*;

public class QuoteClient {
  String quoteServer = "rmi://localhost/quotes";

  public static void main(String[] args) {
```

```
    int num = 0;
    if (args.length > 0) {
        try {
            num = Integer.parseInt(args[0]);
            }
            catch (Exception e) {
            System.out.println("please enter a number as an argument");
            System.exit(0);
    }
  }
  QuoteClient q = new QuoteClient();
  if (num < 2)
        q.getQuote();
  else if (num > 10) // enforcing a limit of 10
        q.getQuotes(10);
  else
        q.getQuotes(num);
  }

  public void getQuote() {
     try {
        RandomQuoter rq =(RandomQuoter)Naming.lookup(quoteServer);
        System.out.println(rq.getQuote());
  } catch (Exception e){System.out.println("ERROR: " + e);}
  }
  public void getQuotes(int n) {
        try {
            RandomQuote rq =(RandomQuote)Naming.lookup(quoteServer);
        System.out.println(rq.getQuotes(n));
     } catch (Exception e){System.out.println("ERROR: " + e);}
  }
}
```

Again, this looks like a normal Java program. Note that you will have to set the domain of your RMI object. We are using "localhost" here. The interesting part is that we are getting a reference to the remote object and casting it to the object that we expect in the line:

```
RandomQuote rq = (RandomQuote)Naming.lookup(quoteServer);
```

We have written 4 classes:

1. The main interface (RandomQuote)
2. The server (QuoteServer)
3. The server implementation (QuoteServerImpl)
4. The client (QuoteClient)

Place these four .java files in a directory by themselves and we will begin to compile them. After each round of compilation, check the directory to see the "magic" that occurs.

Compile the QuoteClient normally: **javac QuoteClient.java**

Next, compile the QuoteServer: **javac QuoteServer.java**

Since QuoteServer requires both RandomQuote and QuoteServerImpl, they will be compiled automatically. If you look in your directory you will see that you now have 4 java files and 4 class files.

Now we need to generate the Stub and Skeleton. This is simple but there is one very important point. The classes need to be in the CLASSPATH to be found by the rmiregistry.

If you simply Type: **rmic QuoteServerImpl** - you will need the location of the QuoteServerImpl in your CLASSPATH.

You can type: **rmic -classpath /your/path/to/ QuoteServerImpl** to specify the path. (*Note that this is four separate elements.*)

Look in your directory again and you will see that the stub and skeleton have been created. You now have the following six classes:

1. QuoteServer.class
2. QuoteServerImpl.class
3. QuoteServerImpl_Skel.class
4. QuoteServerImpl_Stub.class
5. RandomQuote.class
6. QuoteClient.class

Now you will need some "quote" resources. Note that the QuoteServerImpl reads a file from the filesystem and parses it line by line. Any text will work, just make sure to provide a bunch of single lines (more then 10). Make sure that your path is correct.

We can now run the program. This is a three step process. We need to start the registry and the QuoteServer.

Type: **rmiregistry**

then open another shell and...

Type: **java QuoteServer**

You now have two processes going. The QuoteServer console should print the message that we supplied if all is well: *"QuoteServer started and awaiting connections."*

Now open a third shell and type: **java QuoteClient**

You can type an optional number as an argument to get up to 10 random quotes. Notice that the server prints out a message each time a request is handled.

SUMMARY

This chapter is dedicated to java library classes, we have explain major classes of java library which are used commonly.

- The two major classes of strings are **String** and String Buffer class, String Buffer classes used to handle the mutable strings, which can be change.
- File handling is performed by the input and output stream classes.
- Java was designed from the start with networking in mind thats why java provide many features for Networking via java programming.
- Event handling is the main advantage of Java for GUI programming we have explained main event handling examples.

SOLVED PROBLEMS

Q.7.1. Why are streams used in Java?

Ans. In Java streams are used to process input and output data as a sequence of bytes, which does not assume a specific character content or encoding. Byte content can be read from network sources, files and other sources, and bytes copied between multiple inputs and outputs.

Streams types can be sub-classed to add filtering, to mark a point in the stream and re-read those bytes, or to skip a number of bytes. Streams also enable serlializable objects to stored and re-constructed using ObjectInputStream and ObjectOutputStream types.

Q.7.2. What's the difference between a file and a filename? Which is more closely related to the java.io.File class?

Ans. A file is the actual data located at a particular location on a disk. A filename is used to refer to the file. The name of the file can be changed without changing the data. The java.io.File class is closer to a filename than a file

Q.7.3. Explain RMI mechanism?

Ans. Sometimes it is necessary to invoke a method of an object that is being executed on a remote machine as part of distributed computing. This shares computation between 2 machines. Fortunately java has provided this mechanism. This mechanism is called RMI.

Q.7.4. What is the difference between URLinstance and URLConnection instance?

Ans. A URLinstance represents the location of a resource, and a URLConnection instance represents a link for accessing or communicating with the resource at the location.

Q.7.5. How do I make a connection to URL?

Ans. You obtain a URL instance and then invoke openConnection on it.

URLConnection is an abstract class, which means you can't directly create instances of it using a constructor. We have to invoke openConnection method on a URL instance, to get the right kind of connection for your URL.

Eg.
```
URL url;
URLConnection connection;
try{ url = new URL("...");
conection = url.openConnection();
}catch (MalFormedURLException e) { }
```

Q.7.6. What Is a Socket?

Ans. A socket is one end-point of a two-way communication link between two programs running on the network. A socket is bound to a port number so that the TCP layer can identify the application that data is destined to be sent.Socket classes are used to represent the connection between a client program and a server program. The java.net package provides two classes—Socket and ServerSocket—which implement the client side of the connection and the server side of the connection, respectively.

Q.7.7. What information is needed to create a TCP Socket?

Ans. The Local System's IP Address and Port Number. And the Remote System's IPAddress and Port Number.

Q.7.8. What are the two important TCP Socket classes?

Ans. Socket and ServerSocket.

ServerSocket is used for normal two-way socket communication. Socket class allows us to read and write through the sockets. getInputStream() and getOutputStream() are the two methods available in Socket class.

Q.7.9. When MalformedURLException and UnknownHostException throws?

Ans. When the specified URL is not connected then the URL throw MalformedURLException and If InetAddress' methods getByName and getLocalHost are unabletoresolve the host name they throwan UnknownHostException.

Q.7.10. The API doesn't list any constructors for InetAddress- How can we create an InetAddress instance?

Ans. In case of InetAddress the three methods getLocalHost, getByName, getByAllName can be used to create instances.

E.g.

```
InetAddress add1;
InetAddress add2;
try{
add1 = InetAddress.getByName("java.sun.com");
add2 = InetAddress.getByName("199.22.22.22");
}catch(UnknownHostException e){}
```

Q.7.11. Is it possible to get the Local host IP?

Ans. Yes. Use InetAddress's getLocalHost method.

Q.7.12. What's the Factory Method?

Ans. Factory methods are merely a convention whereby static methods in a class return an instance of that class. The InetAddress class has no visible constructors. To create an InetAddress object, you have to use one of the available factory methods. In InetAddress the three methods getLocalHost, getByName, getByAllName can be used to create instances of InetAddress.

Q.7.13. What's the difference between TCP and UDP?

Ans. These two protocols differ in the way they carry out the action of communicating. A TCP protocol establishes a two way connection between a pair of computers, while the UDP protocol is a one-way message sender. The common analogy is that TCP is like making a phone call and carrying on a two-way communication, while UDP is like mailing a letter.

Q.7.14. What is the Proxy Server?

Ans. A proxy server speaks the client side of a protocol to another server. This is often required when clients have certain restrictions on which servers they can connect to. And when several users are hitting a popular web site, a proxy server can get the contents of the web server's popular pages once, saving expensive internetwork transfers while providing faster access to those pages to the clients.

Also, we can get multiple connections for a single server.

Q.7.15. What are the seven layers of OSI model?

Ans. Application, Presentation, Session, Transport, Network, DataLink, Physical Layer.

Q.7.16. What Transport Layer does?

Ans. It ensures that the mail gets to its destination. If a packet fails to get its destination, it handles the process of notifying the sender and requesting that another packet be sent.

Q.7.17. What is DHCP?

Ans. Dynamic Host Configuration Protocol, a piece of the TCP/IP protocol suite that handles the automatic assignment of IP addresses to clients.

Q.7.18. What is SMTP?

Ans. Simple Mail Transmission Protocol, the TCP/IP Standard for Internet mails. SMTP exchanges mail between servers; contrast this with POP, which transmits mail between a server and a client.

Q.7.19. Write a program to recursively list all the files and directories contained in a directory given on the command line.

Ans.

```
import java.io.File;

public class rls {

  File dir;

  public static void main(String[] args) {

    for (int i = 0; i < args.length; i++) {
    rls r = new rls(args[i]);
    r.list();
    }

  }

  public rls(String s) {

    dir = new File(s);
  }

    // This method doesn't do anything if dir isn't a directory
    public void list() {

    if (dir.isDirectory()) {
    System.out.println("\n" + dir + ":\n");
    String[] files = dir.list();
    for (int i = 0; i < files.length; i++) {
    System.out.println(files[i]);
    }
    for (int i = 0; i < files.length; i++) {
    rls r = new rls(dir + File.separator + files[i]);
    r.list();
    }

  }

  }

}
```

Note: This program can get stuck in an infinite loop if aliases or symbolic links point to a directory that's already been scanned. How would you fix this?

Q.7.20. Difference between = = and eauals method?

Ans. = = compares the references(hash codes) of the strings so it gives unreliable results. Equals compares the contents of the strings.

EXERCISES

Q.7.1. How can you create a string in Java programming?

Q.7.2. What do you mean by immutable strings?

Q.7.3. How can we compare two strings? Give an example.

Q.7.4. What are the difference between string class Vs String Buffer Class.

Q.7.5. Explain the use of FileReader Class.

Q.7.6. Explain the use of following Networking classes:

(*a*) URL

(*b*) InetAddress

(*c*) ServerSocket

Q.7.7. Explain the concept of RMI?

Q.7.8. What do you mean by Events?

Q.7.9. What are the events we can handle by the programming?

Q.7.10. What are the information needed for TCP sockets?

Chapter 8

JAVA AWT

8.1. INTRODUCTION

The AWT stands for Abstract Window ToolKit. It is a portable GUI library between Solaris and Windows 95/NT and Mac System 7.X (soon) for stand-alone applications and/or applets. Since it can be used in applets it can be used on IRIX, SunOS, HP/UX, Linux which Netscape 2.0 supports.

The Abstract Window Toolkit provides many classes for programmers to use. It is your connection between your application and the native GUI. The AWT hides you from the underlying details of the GUI your application will be running on and thus is at very high level of abstraction. It takes the lowest common denominator approach to retain portability. No floating toolbars or Balloon help here.

The Abstract Window Toolkit is a Java package and can be used in any Java program by importing java.awt.* via the import keyword.

The Abstract Window Toolkit contains three kind of classes

- Containers
  – Frame, Dialog, Panel, Applet etc.
- Components
  – Label, TextField, Button, Checkbox, Scrollbar etc.
- Supporting classes
  – Color, Font, Dimesion etc.

The Abstract Window Toolkit provides a class Component which an abstract class that provides all common methods to all the containers and components

CHAPTER OUTLINE

- Introduction
- Working with window
- AWT Controls
  - Buttons
  - Canvas
  - Checkbox
  - CheckboxGroup
  - Choice
  - Label
  - List
  - Scrollbar
  - TextField
  - TextArea
- Layout Manager
  - BorderLayout
  - CardLayout
  - GridLayout
  - GridBagLayout
  - FlowLayout
- Menus
- Images
- Additional Packages
- Java Utility Package
  - Java language Package
  - Java collections Package

> Concept
>
> *AWT is a library of classes which provides GUI tools to develope GUI applications and applets.*

- void setSize(int width, int height)
- void setLocation(int x, int y)
- void setBounds(int x, int y, int w, int h)
- void setVisible(boolean b)
- void setBackground(Color c)
- void setFont(Font f)
- void setCursor(Cursor c)

GUI programs can be developed in two ways

1. Java Applications
   - Loaded on every machine
   - To make a java program as GUI application, it must inherit from Frame class
2. Java Applets
   - Run using a web browser
   - Place only on one machine called web server
   - To make a program as applet, it must inherit from Applet class

8.2. WORKING WITH WINDOWS

The first thing you need to do when creating a GUI application is create the window. In Java AWT, top-level windows are represented by the Frame class.

Example 8.1: Program to create a Window in Java Application.

```
import java.awt.*;
class AWTexample extends Frame
{
    public AWTexample()
    {
        setTitle("My Window");
        setSize(300,200);
    }
    public static void main( String arg[])
    {
    AWTexample F=new AWTexample();
    F.setVisible(true);
    }
}
```

COMPILE:

```
C:\>javac AWTexample.java
```

EXECUTE:

```
C:\>java AWTexample.java
```

OUTPUT

> ✍ **Note:** The output window will not close using close button on title bar because it needs event handling, which will discuss in next chapter. Use Ctrl+C on command window to close.

Example 8.2: Program to create a Window in Java Application using Frame object.

```
import java.awt.*;
class AWTexample
{
    private Frame F;
    public AWTexample()
    {
        F=new Frame();
        F.setTitle("My Window");
        F.setSize(300,200);
        F.setBackground(Color.blue);
        F.setVisible(true);
    }
public static void main( String arg[])
    {
    new AWTexample();
    }
}
```

COMPILE:

```
C:\>javac AWTexample.java
```

EXECUTE:

```
C:\>java AWTexample.java
```

OUTPUT

A Panel, like a Frame, provides the space for you to attach any GUI components, including others panels. Each Panel, which inherits from Container, can have its own layout manager.

After you create a Panel, you must add it to a Window or Frame. This is done using the add method of the Container class. Then set the frame to visible so that the frame and panel are displayed.

Example 8.3: Program to create a small yellow Panel, and adds it to a Frame.

```
import java.awt.*;
class AWTexample
{
    private Frame F;
    private Panel pan;
    public AWTexample()
    {
        F = new Frame();
        pan = new Panel();
    }
    public void show()
    {
        F.setTitle("My Window");
        F.setSize(300,200);
        F.setBackground(Color.blue);
        F.setLayout(null);
        pan.setSize(100,100);
        pan.setBackground(Color.yellow);
        F.add(pan);
        F.setVisible(true);
    }

public static void main( String arg[])
    {
        AWTexample A = new AWTexample();
```

```
            A.show();
        }
    }
```

COMPILE:

```
C:\>javac AWTexample.java
```

EXECUTE:

```
C:\>java AWTexample.java
```

OUTPUT

8.3. AWT CONTROLS

Controls are components that allow a user to interact with your application. The AWT supports the following types of controls:

- Buttons
- Checkbox
- CheckboxGroup
- Choice
- Label
- List
- Scrollbar
- TextField
- TextArea

Next, examine each of the AWT components below. Most, but not all, directly extend Component.

You've used most of the components should be familiar to you.

8.3.1. Button

A Button has a single line label and may be "pushed" with a mouse click. See an example how to add a button on the window.

Example 8.4: Program to add a Button on window.

```
import java.awt.*;
class AWTexample extends Frame
```

```
{
    public AWTexample()
    {
        setTitle("My Window");
        setSize(300,200);
        Button B=new Button("Press Me");
        setLayout(null);
        B.setBounds(50,50,90,30);
        add(B);
        setVisible(true);
    }

public static void main( String arg[])
    {
        new AWTexample();
    }
}
```

COMPILE:

```
C:\>javac AWTexample.java
```

EXECUTE:

```
C:\>java AWTexample.java
```

OUTPUT

Note: In the above example there is no event handling added; pressing the button will not do anything.

The AWT button has no direct support for images as labels.

8.3.2. CheckBox

A Checkbox is a label with a small pushbutton. The state of a Checkbox is either true (button is checked) or false (button not checked). The default initial state is false. Clicking a Checkbox toggles its state. For example:

Example 8.5: Program to add a Check-Box on window.

```
import java.awt.*;
class AWTexample extends Frame
```

```
{
    public AWTexample()
    {
        setTitle("My Window");
        setSize(300,200);
        Checkbox C=new Checkbox("Check Me");
        setLayout(null);
        C.setBounds(50,50,90,30);
        add(C);
        setVisible(true);
    }
public static void main( String arg[])
{
    new AWTexample();
}
```

COMPILE:

```
C:\>javac AWTexample.java
```

EXECUTE:

```
C:\>java AWTexample.java
```

OUTPUT

8.3.3. CheckBoxGroup

A CheckboxGroup is used to control the behavior of a group of Checkbox objects (each of which has a true or false state). Exactly one of the Checkbox objects is allowed to be true at one time. Checkbox objects controlled with a CheckboxGroup are usually referred to as "radio buttons".

The following example illustrates the basic idea behind radio buttons.

Example 8.6: Program to add two Radio Buttons on window.

```
import java.awt.*;
class AWTexample extends Frame
{
    public AWTexample()
```

```
        {
            setTitle("My Window");
            setSize(300,200);
            CheckboxGroup cbg=new CheckboxGroup();
            Checkbox C1=new Checkbox("I am Indian", cbg, ture);
            Checkbox C2=new Checkbox("I am American", cbg, false);
            setLayout(null);
            C1.setBounds(50,50,90,30);
            C2.setBounds(50,70,100,30);
            add(C1);
            add(C2);
            setVisible(true);
        }
        public static void main( String arg[])
        {
            new AWTexample();
        }
    }
```

COMPILE:

C:\>javac AWTexample.java

EXECUTE:

C:\>java AWTexample

OUTPUT

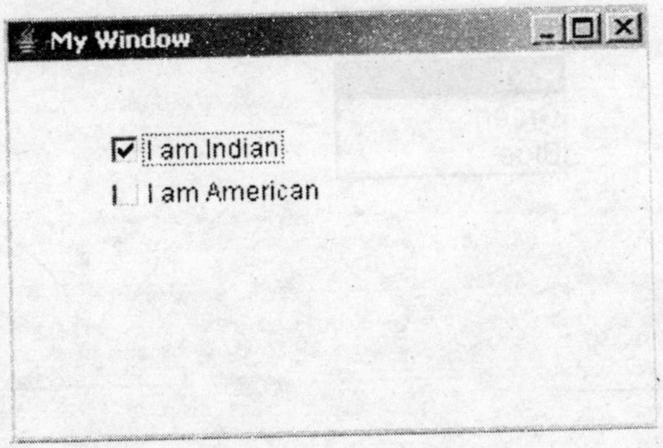

8.3.4. Choice

Choice objects are drop-down lists. The visible label of the Choice object is the currently selected entry of the Choice.

Example 8.7: Program to add a Combo-Box on window.

```
import java.awt.*;
class AWTexample extends Frame
{
    public AWTexample()
    {
        setTitle("My Window");
```

```
            setSize(300,200);
            Choice C=new Choice();
            setLayout(null);
            C.setBounds(50,50,90,30);
            C.add("Red");
            C.add("Green");
            C.add("Blue");
            add(C);
            setVisible(true);
        }
        public static void main( String arg[])
        {
            new AWTexample();
        }
    }
```

COMPILE:

```
C:\>javac AWTexample.java
```

EXECUTE:

```
C:\>java AWTexample
```

OUTPUT

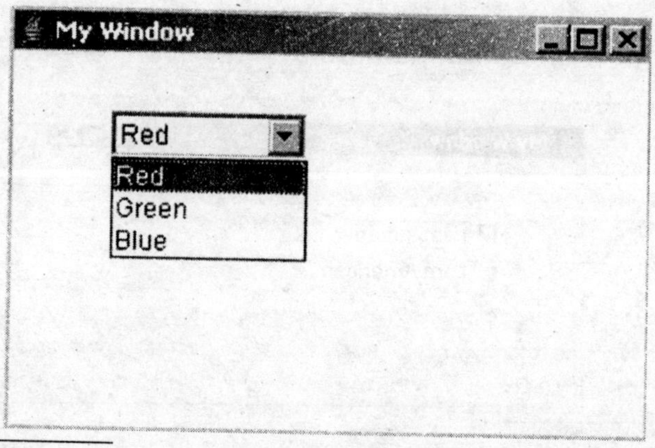

8.3.5. Label

A Label is a displayed Label object. It is usually used to help indicate what other parts of the GUI do, such as the purpose of a neighboring text field.

Example 8.8: Program to add two Labels on window.

```
import java.awt.*;
class AWTexample extends Frame
{
    public AWTexample()
    {
        setTitle("My Window");
        setSize(300,200);
```

```
            Label L1=new Label(" Label-1");
            Label L2=new Label(" Label-2");
            setLayout(null);
            L1.setBounds(50,50,90,30);
            L2.setBounds(50,70,100,30);
            add(L1);
            add(L2);
            setVisible(true);
        }

        public static void main( String arg[])
        {
            new AWTexample();
        }
    }
```

COMPILE:

```
C:\>javac AWTexample.java
```

EXECUTE:

```
C:\>java AWTexample
```

OUTPUT

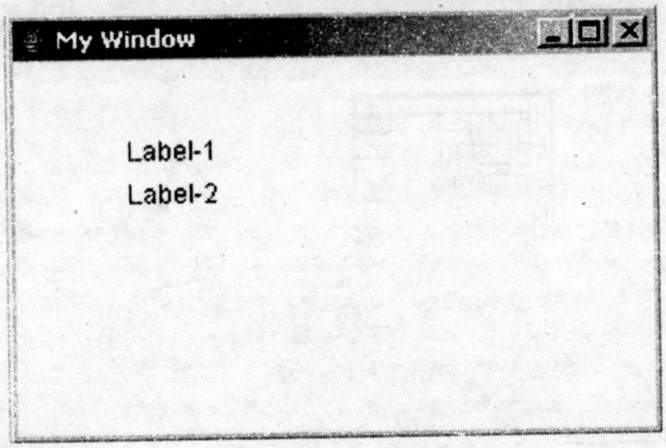

8.3.6. List

A List is a scrolling list box that allows you to select one or more items. Multiple selections may be used by passing true as the second argument to the constructor.

Example 8.9: Program to add a List-Box on window.

```
import java.awt.*;
class AWTexample extends Frame
{
    public AWTexample()
    {
        setTitle("My Window");
        setSize(300,200);
        List C=new List(4,false);
```

```
            setLayout(null);
            C.setBounds(50,50,90,50);
            C.add("India");
            C.add("Pakistan");
            C.add("Sri Lanka");
            C.add("Canada");
            add(C);
            setVisible(true);
        }
    public static void main( String arg[])
        {
            new AWTexample();
        }
    }
```

COMPILE:

```
C:\>javac AWTexample.java
```

EXECUTE:

```
C:\>java AWTexample
```

OUTPUT

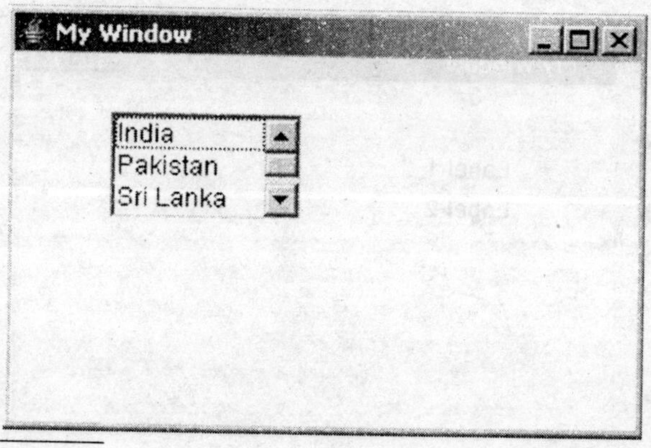

8.3.7. Scroll-Bar

A Scrollbar is a "slider" widget with characteristics specified by integer values that are set during Scrollbar construction. Both horizontal and vertical sliders are available.

Example 8.10: Program to add a Scroll-Bar on window.

```
import java.awt.*;
class AWTexample extends Frame
{
    public AWTexample()
    {
        setTitle("My Window");
        setSize(300,200);
        Scrollbar C=new Scrollbar(Scrollbar.HORIZONTAL,0,5,-100,105);
```

```
        //Scrollbar(HoR/VER, intial value, width of slider, min rage, max range)
        setLayout(null);
        C.setBounds(50,50,150,30);
        add(C);
        setVisible(true);
    }
    public static void main( String arg[])
    {
        new AWTexample();
    }
}
```

COMPILE:

```
C:\>javac AWTexample.java
```

EXECUTE:

```
C:\>java AWTexample
```

OUTPUT

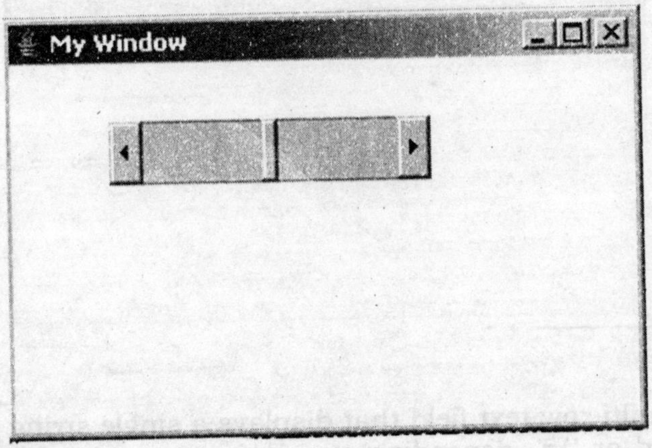

8.3.8. TextField

A TextField is a scrollable text display object with one row of characters. The preferred width of the field may be specified during construction and an initial string may be specified.

Example 8.11: Program to add a Text-Field on window.

```
import java.awt.*;
class AWTexample extends Frame
{
    public AWTexample()
    {
        setTitle("My Window");
        setSize(300,200);
        TextField C=new TextField("Type Some thing");
        setLayout(null);
        C.setBounds(50,50,100,30);
```

```
            add(C);
            setVisible(true);
        }
    public static void main( String arg[])
        {
            new AWTexample();
        }
    }
```

COMPILE:

```
C:\>javac AWTexample.java
```

EXECUTE:

```
C:\>java AWTexample
```

OUTPUT

8.3.9. TextArea

A TextArea is a multi-row text field that displays a single string of characters, where newline ('\n' or '\n\r' or '\r', depending on platform) ends each row. The width and height of the field is set at construction, but the text can be scrolled up/down and left/right.

Example 8.12: Program to add a Text-Area on window.

```
import java.awt.*;
class AWTexample extends Frame
{
    public AWTexample()
    {
        setTitle("My Window");
        setSize(300,200);
        TextArea C=new TextArea("Type code hare", 10,30);
        setLayout(null);
        C.setBounds(20,20,250,180);
        add(C);
        setVisible(true);
```

```
    }
    public static void main( String arg[])
    {
        new AWTexample();
    }
}
```

COMPILE:

```
C:\>javac AWTexample.java
```

EXECUTE:

```
C:\>java AWTexample
```

OUTPUT

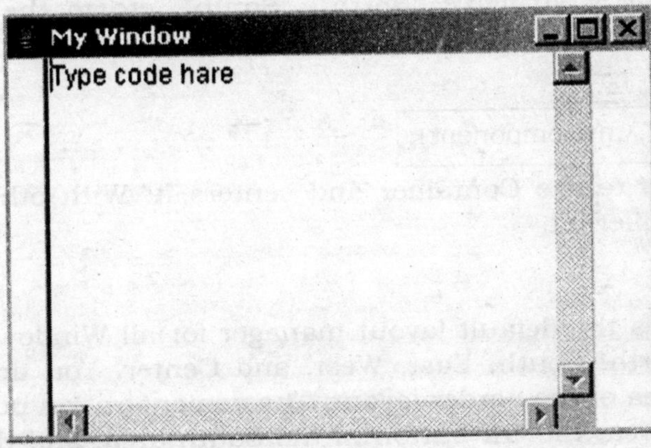

There is no way, for example, to put the cursor at beginning of row five, only to put the cursor at single dimension position 50.

There is a four-argument constructor that accepts a fourth parameter of a scrollbar policy. The different settings are the class constants: SCROLLBARS_BOTH, SCROLLBARS_HORIZONTAL_ONLY, SCROLLBARS_NONE, SCROLLBARS_VERTICAL_ONLY. When the horizontal (bottom) scrollbar is not present, the text will wrap.

8.4. LAYOUT MANAGER

How Components are "laid out" within a Container is described by the LayoutManager class. Since the LayoutManager class is abstract, we can not use it directly. You must sub-class it and provide your own functionality or use a *derived class of LayoutManager*(i.e BorderLayout, CardLayout, GridLayout, etc) already created for you.

There are many different layout schemes, but the ones pre-defined for us are

- **BorderLayout** This scheme lays out the Component in 5 ways
  1. North-Northern part of the Container
  2. South-Southern part of the Container
  3. East-Eastern part of the Container
  4. West-Western part of the Container
  5. Center-Centered in the Container

- **CardLayout** - Allows for what Windows programmers have called for years "tabbed dialogs" or dynamic dialogs(now available on all versions of Netscape).
- **GridLayout** - Allows for the layout of Components in a grid -like fashion rather than "North" or "Center".
- **GridBagLayout** - HTMLTable-ish style of layout
- **FlowLayout** - Allows for Component to be laid out in a row(or flow) and aligned(left, right, center).
- **None**-No layout, the Container will not attempt to reposition the Components during a update.

To use a layout we must call setLayout() for the Container with an instance of a LayoutManager(since LayoutManager is abstract, a sub-class will do).

Using each layouts is different. For BorderLayout you specify the type of layout(North, South, etc.) by passing a string(i.e "North", "South", etc to the method add() in a Container.

i.e

```
add("Center", myComponent);
```

adds *myComponent* to the Container and centers it. With other layouts, the first **String** parameter is different.

8.4.1. BorderLayout

The border layout is the default layout manager for all Window objects. It consists of five fixed areas: North, South, East, West, and Center. You do not have to put a component in every area of the border layout. The demonstration puts the label "Border Layout" in the North area, the OK button in the South area, and the List of fruit trees in the Center area. The East (right) and West (left) areas are empty.

The five areas of a border layout are shown at left. Notice how the North and South areas run like a header and footer across the panel.

If any or all of the North, South, East, or West areas are left out, the Central area spreads into the missing area or areas. However, if the Central area is left out, the North, South, East, or West areas do not change.

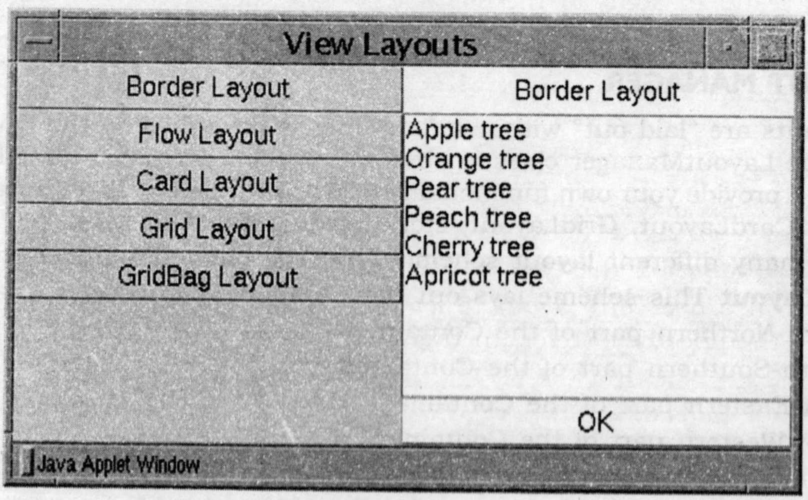

Fig. 8.8. Border layout

You can indent the Central area by placing a label with one or more blank characters in the East or West areas. The following code shows two ways to do it:

```
Label label1 = new Label(" ");
p2.add("East, label1);

Label label2 = new Label();
label2.setText(" ");
p2.add("East", label2);
```

Here is the code for the border layout panel with the fruit tree list shown above:

```
p2.setLayout(new BorderLayout());
Label dialogLabel = new Label();
dialogLabel.setAlignment(Label.CENTER);
dialogLabel.setText("Border Layout");

List items = new List(6, true);
items.addItem("Apple tree");
items.addItem("Orange tree");
items.addItem("Pear tree");
items.addItem("Peach tree");
items.addItem("Cherry tree");
items.addItem("Apricot tree");

OK = new Button("OK");
p2.add("North", dialogLabel);
p2.add("South", OK);
p2.add("Center", items);
```

Note that when using the BorderLayout, you must add components with either the add(String, Component) or the add(Component, Object) method to ensure the components are all added correctly. A common mistake people make is to use add(Component), which can result in some of the components not being visible.

8.4.2. CardLayout

The card layout lets you use one container (usually a panel) to display one out of many possible component children (like flipping cards on a table). A program can use this layout to show a different child component to different users. For example, the interface shown to an administrator might have additional functionality from the interface shown to a regular user. With card layout, your program can show the appropriate interface depending on the type of user using the program.

You can also use card layout to let your end user toggle among different displays and choose the one they prefer. In this case, the program must provide a GUI for the user to make the selection. The Choice component is commonly used for this case and the following card layout example shows how to use a Choice component to set up a card layout that lets the end user toggle between two interfaces.

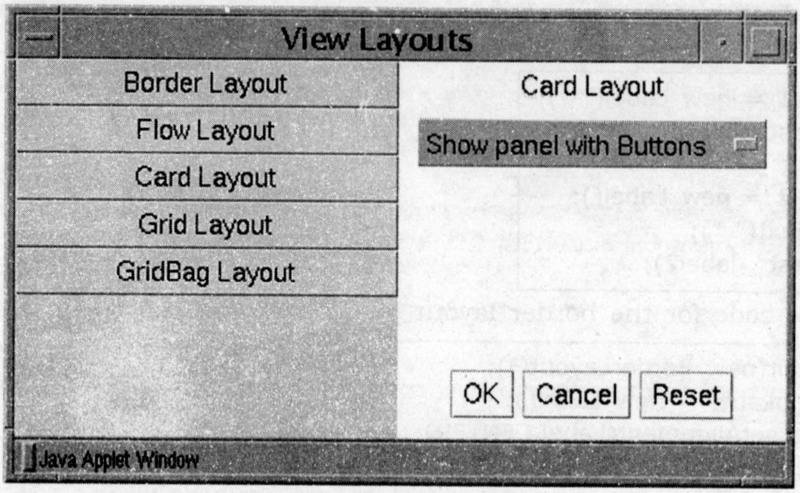

Fig. 8.16. Card layout

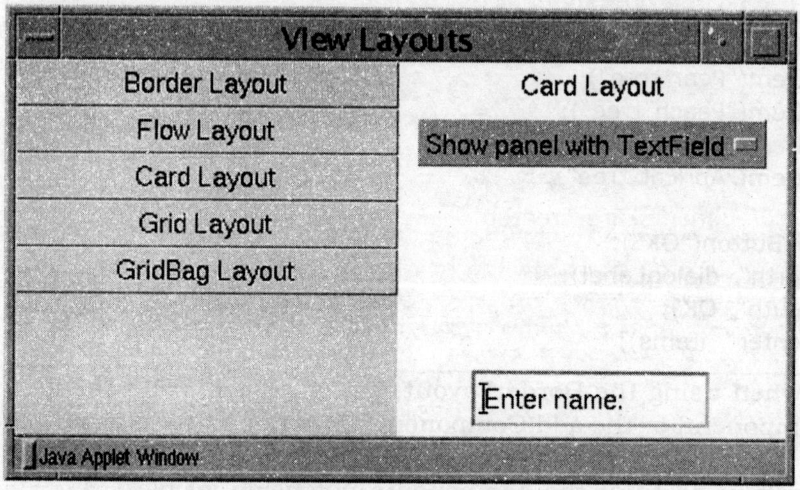

Fig. 8.17. Card layout

Here is the code for the card layout panel shown above:

```
String BUTTONPANEL = "Show panel with Buttons";
String TEXTPANEL = "Show panel with TextField";

//Border layout is used to set the right panel up
//with the Choice component
//and "Card Layout" label

p2.setLayout(new BorderLayout());

//The Choice component is put in a panel for
//better appearance
Panel c2 = new Panel();
Choice c = new Choice();
```

```
c.addItem(BUTTONPANEL);
c.addItem(TEXTPANEL);
c2.add(c);

//The Label is centered and put in the North area
//of the right panel
Label cardLabel = new Label("Card Layout");
cardLabel.setAlignment(Label.CENTER);
p2.add("North", cardLabel);

//The choice is added to the
//Center area of the right panel
p2.add("Center", c2);

//The card panel is created and CardLayout used
cards = new Panel();
cards.setLayout(new CardLayout());

//The two panels controlled by the card panel
//are created.
p3 = new Panel();
p4 = new Panel();

//The button components are created
//and added to p3
OK = new Button("OK");
Button quit = new Button("Cancel");
reset = new Button("Reset");
p3.add(OK);
p3.add(quit);
p3.add(reset);

//The text field component is
//created and added to p4
tfield = new TextField("Enter name:", 20);
p4.add(tfield);

//Panels p3 and p3 are added to
//the card panel./c
cards.add(BUTTONPANEL, p3);
cards.add(TEXTPANEL, p4);

//The card panel is added to the
//south area of the right panel
p2.add("South", cards);
```

8.4.3. GridLayout

The Grid layout arranges components into a grid of rows and columns. You specify the number of rows and columns, the number of rows only and let the layout manager determine the numer of columns, or the number of columns only and let the layout manager determine the number of rows.

The cells in the grid are equal size based on the largest component in the grid. Resize the window to see how the components are resized to fit cells as they get larger and smaller.

The grid layout demonstration starts with a one column grid, and lets you choose to see the same components in two, three, and four column grids.

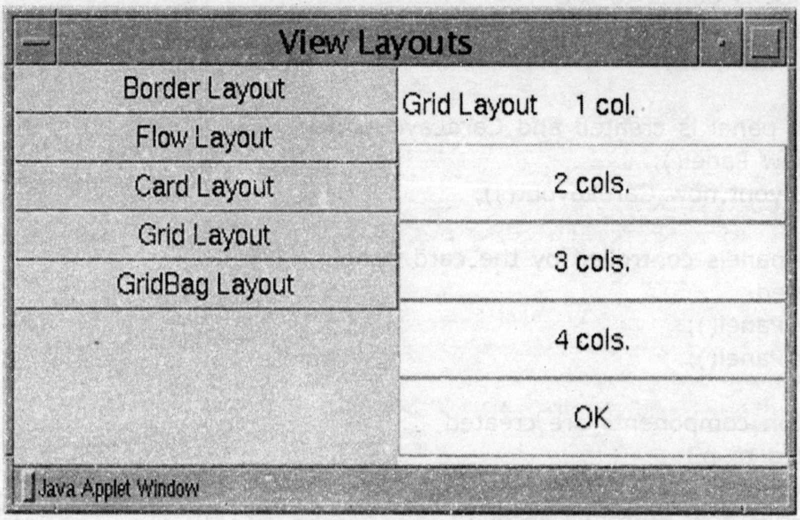

Fig. 8.18. Grid layout

Here is the code for the grid layout panel shown above.

```
    //Set grid to 5 rows and 1 column
p2.setLayout(new GridLayout(5,1));

two = new Button(“2 cols.”);
three = new Button(“3 cols.”);
four = new Button(“4 cols.”);
OK = new Button(“OK”);

p2.add(new Label(“Grid Layout 1 col.”));

p2.add(two);
p2.add(three);
p2.add(four);
p2.add(OK);

//Implement 2 column grid.
//Let layout manager determine number of rows.
p2.setLayout(new GridLayout(0,2));
```

```
two = new Button("2 cols.");
three = new Button("3 cols.");
four = new Button("4 cols.");
OK = new Button("OK");
p2.add(two);
p2.add(three);
p2.add(four);
p2.add(OK);

//Implement 3 column grid.
//Let layout manager determine number of rows.
p2.setLayout(new GridLayout(0,3));
two = new Button("2 cols.");
three = new Button("3 cols.");
four = new Button("4 cols.");
OK = new Button("OK");
p2.add(two);
p2.add(three);
p2.add(four);
p2.add(OK);

//Implement 4 column grid.
//Let layout manager determine number of rows.
p2.setLayout(new GridLayout(0,4));
two = new Button("2 cols.");
three = new Button("3 cols.");
four = new Button("4 cols.");
OK = new Button("OK");
p2.add(two);
p2.add(three);
p2.add(four);
p2.add(OK);
```

8.4.4. GridBagLayout

The Grid bag layout (like grid layout) arranges components into a grid of rows and columns, but lets you specify a number of settings to fine-tune how the components are sized and positioned within the cells. Unlike the grid layout, the rows and columns are not constrained to be a uniform size. For example, a component can be set to span multiple rows or columns, or you can change its position on the grid.

The layout manager uses the number of components in the longest row, the number of rows created, and the size of the components to determine the number and size of the cells in the grid. The set of cells a component spans can be referred to as its display area. A component's display area and the way in which it fills that display area are defined by a set of constraints that are represented by the GridBagConstraints object. These constraints are given below:

- gridx / gridy
- gridwidth/gridheight

- fill
- anchor
- ipadx / ipady
- insets

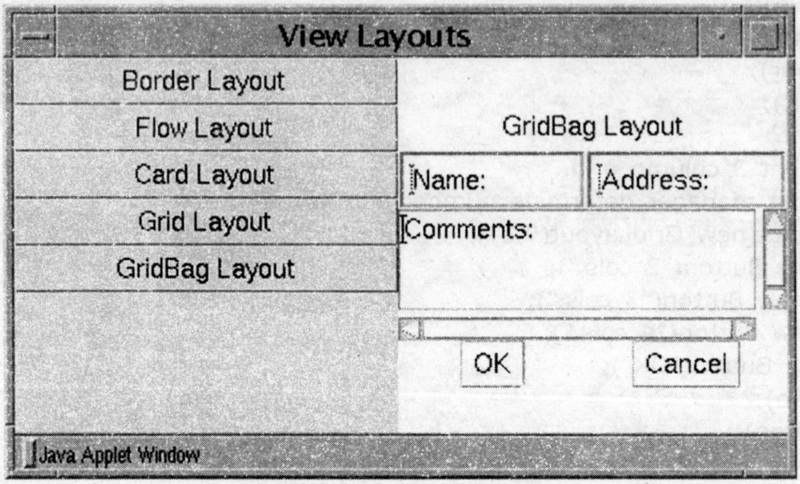

Fig. 8.19. Grid Bag Layout

The GridBagConstraints object is initialized with a set of default values, each of which can be customized to change the way the component will be layed out within the gridbag.

The following bulleted list outlines how to use the GridBagConstraints and GridBagLayout objects in code to achieve the look you want. This outline is for instructional purposes only and is used in the following code example. Once you understand how the objects work together, your code can, of course, do things in a different order or take another approach.

- Decide how you want the components sized and positioned.
- Create a GridBagLayout and GridBagConstraints object.
- Have the container use the GridBagLayout object.
- Customize the GridBagConstraints settings (or use defaults).
- Create Component A.
- Pass Component A and GridBagConstraints to the GridBagLayout object.
- Add Component A to the container. The previous bullet told the container's layout which settings to use for the component.
- Create Component B.
- Change the GridBagConstraints settings or leave as is.
- Pass Component B and GridBagConstraints to the GridBagLayout object.
- Add Component B to the container.
- and so on . . . .

Here is the code for the grid bag layout shown above.

```
//Create GridBagLayout and
//GridBagConstraints objects.
 //Have container p2 use grid bag layout.
```

```
GridBagConstraints c = new GridBagConstraints();
GridBagLayout gridbag = new GridBagLayout();
p2.setLayout(gridbag);

//Make components fill the available display area
//in the horizontal and vertical directions
c.fill = GridBagConstraints.BOTH;
//Resize columns and keep them equal
//during window resizing
//Note: Since no weighty value is
//specified, rows are not resized,
//but remain together in the middle of the
//display.
//Test this by resizing the demonstration
//window with GridBag Layout selected.
c.weightx = 1.0;

//Create the Label component.
Label top = new Label("GridBag Layout");
top.setAlignment(Label.CENTER);

//Make the Label
component the last one in its row.
c.gridwidth = GridBagConstraints.REMAINDER;

//Pass the Label component and
//GridBagConstraints object to layout.
gridbag.setConstraints(top, c);

//Add the Label component to container p2.
//The GridBagConstraints
//passed to the GridBagLayout object for
//this container put the
//label in a row by itself, size the label
//to fill the row in the
//horizontal and vertical directions, and
//cause columns to resize
//equally when the window is resized.
p2.add(top);

//Create a Text Field component.
TextField name = new TextField("Name:", 25);

//Make the Text Field component the next
//to last one in its row.
//This means 1 or more components can be added
//to the row after
```

```
//the Text field component.
c.gridwidth = GridBagConstraints.RELATIVE;

//Pass the Text Field component and
//GridBagConstraints object to layout.
gridbag.setConstraints(name, c);

//Add the Text Field component to
//container p2.
p2.add(name);

//Create another Text Field component.
TextField addr = new TextField("Address:", 25);

//Make the Text field component the
//last one in its row.
c.gridwidth = GridBagConstraints.REMAINDER;

//Pass the Text Field component and
//GridBagConstraints object to layout.
gridbag.setConstraints(addr, c);

//Add the Text Field component to container p2.
p2.add(addr);

//Create a Text Area component.
TextArea comments = new TextArea(3, 25);
comments.setEditable(true);
comments.setText("Comments:");

//Pass the Text Area component and
//GridBagConstraints object to layout.
//Because GridBagConstraints are unchanged,
//the Text Area component
//is in a row by itself.
gridbag.setConstraints(comments, c);

//Add the Text Area component to container p2.
p2.add(comments);

//Create the OK and Cancel buttons.
OK = new Button("OK");
cancel = new Button("Cancel");

//Make the OK button the next to last
//one in its row.
//Components should not expand to fill
//the row in either
```

```
//the horizontal or vertical direction.
c.fill = GridBagConstraints.NONE;
c.gridwidth = GridBagConstraints.RELATIVE;

//Pass the OK button and GridBagConstraints
//object to layout.
gridbag.setConstraints(OK, c);

//Make the Cancel button the last one
//in its row.
c.gridwidth = GridBagConstraints.REMAINDER:

//Pass the Cancel button and
//GridBagConstraints object to layout.
gridbag.setConstraints(cancel, c);

//Add the OK and Cancel buttons to
//container p2.
p2.add(OK);
p2.add(cancel);
```

8.4.5. Flow Layout

The flow layout is the default layout manager for all Panel objects and applets. It places the panel components in rows according to the width of the panel and the number and size of the components. The best way to understand flow layout is to resize the demonstration window and notice how the components flow from one row to the other as you make the window wide and narrow.

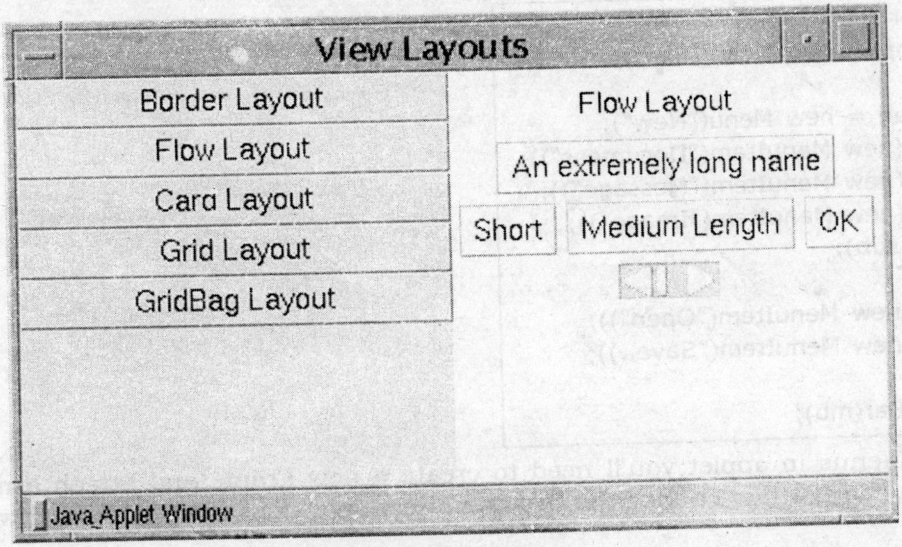

Fig. 8.20. Flow layout

Here is the code for the flow layout panel shown above:

```
        p2.setLayout(new FlowLayout());

    p2.add(new Label("Flow Layout"));
    OK = new Button("OK");
    p2.add(new Button("An extremely long name"));
    p2.add(new Button("Short"));
    p2.add(new Button("Medium Length"));
    p2.add(OK);
    p2.add(new Scrollbar(Scrollbar.HORIZONTAL));
```

8.5. MENUS

Menus are at the center of your application. The difference between a usuable application and one that is absolutely frustrating lies in the organizaticn of your menus. The user-interfaces rules for a good menu design are not clear.

The following classes are related to Menus

- MenuComponent
  - MenuBar
  - MenuItem
    - CheckboxMenuItem
    - Menu

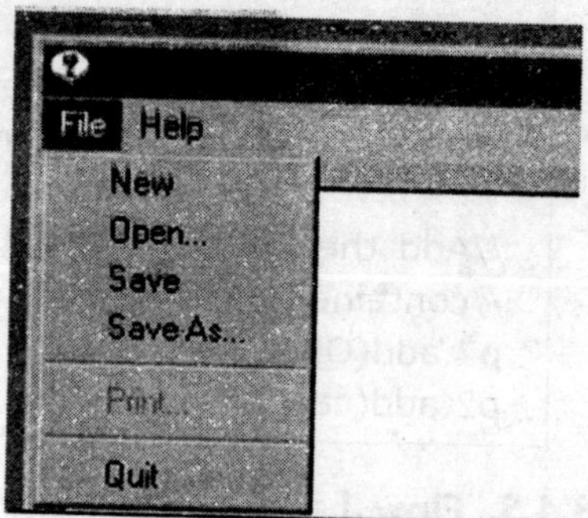

Fig. 8.21. Menus

Handling is done in the handleEvent() method which you will override.

To use menus you need to create a MenuBar and Menus with MenuItems, then attach the MenuBar to your application via setMenuBar. You can even embed, Menus in MenuItems..

```
MenuBar mb = new MenuBar();
Menu top = new Menu("File");

Menu sub = new Menu("New");
sub.add(new MenuItem("Document"));
sub.add(new MenuItem("Message"));
sub.add(new MenuItem("Image"));
top.add(sub);

top.add(new MenuItem("Open"));
top.add(new MenuItem("Save..));

setMenuBar(mb);
```

To use Menus in applet you'll need to create a new Frame and attach a menubar to that. You can not just change the browser's menubar. You are not dealing with OLE 2 here.

8.6. IMAGES

Have you ever wanted to do simple image manipulations like flipping, scaling, and cropping, but didn't want to bother with image filters? Starting with Java 1.1, you can do this very easily with a new drawImage method. The method signature is as follows:

```
public boolean drawImage (Image img, int dx1, int dy1, dx2, dy2,
    int sx1, int sy1, sx2, sy2,[Color backgroundColor,]
    ImageObserver observer)
```

The Image, optional background color, Color, and ImageObserver parameters are shared with the Java 1.0 varieties of the method, along with the return value. However, new and different to Java 1.1 are the four sets of points. They describe source and destination bounding rectangles for the image. Starting with the 's' parameters, the coordinates describe the source piece of the original image to draw. If the (x1, y1) corner is (0, 0), and the (x2, y2) corner is (img.getWidth(this), img.getHeight(this)), then the entire image will be drawn. On the other hand, if the coordinates specify a smaller size, the image is cropped to the area specified. The source corners must be the top-left and bottom-right corners of the area to draw, as shown in figure 8.22.

Fig. 8.22. Image

The following code in a Java 1.1 program demonstrates cropping to generate figure 8.22.

```
g.drawImage (im, 0, 0, 135, 90,65, 5, 200, 95, this);
```

Working with the 'd' parameters is where the real interesting things happen. These parameters describe the destination drawing area for the piece of the source image you selected. Since the points must be opposite corners of a rectangle, four options are available for these points:

- Keeping the coordinates in the same corners
- Swapping x-coordinates
- Swapping y-coordinates
- Swapping both coordinates

Keeping the coordinates in the same corners results in the scaling of the image. (No scaling will be done if the differences between the height and width of the two point sets is identical.) See figure 8.23 below.

Fig. 8.23. Scaled image to half width and twice height

The following source will compress the width to half size and double the height:

```
g.drawImage (im, 0, 0, im.getWidth(this)/2, im.getHeight(this)*2,
    0, 0, im.getWidth(this), im.getHeight(this), this);
```

Swapping x coordinates has the points in the top-right and bottom-left corners. This results in a flip along the vertical axis, as shown in figure 8.24. Here's the code for implementing this flip.

```
g.drawImage (im, im.getWidth(this), 0, 0, im.getHeight(this),
    0, 0, im.getWidth(this), im.getHeight(this), this);
```

Fig. 8.24. Flipping the image along the vertical axis

Swapping y coordinates, as shown in figure 8.25 and in the code below, has the points in the bottom-left and top-right corners. This results in a flip along the horizontal axis.

```
g.drawImage (im, 0, im.getHeight(this), im.getWidth(this), 0,
    0, 0, im.getWidth(this), im.getHeight(this), this);
```

Fig. 8.25. Flipping the image along the horizontal axis

Finally, figure 8.26 shows how swapping both coordinates results in a flip along both axis. The code is as follows:

```
g.drawImage (im, 0, im.getHeight(this), im.getWidth(this), 0,
    0, 0, im.getWidth(this), im.getHeight(this), this);
```

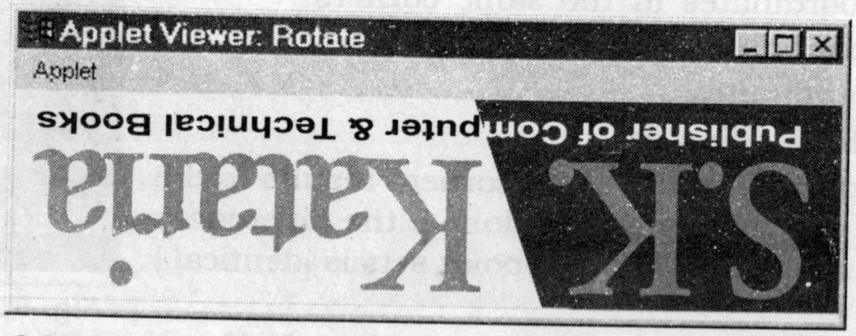

Fig. 8.26. Flipping the image along the vertical and horizontal axes

That's all there is to the new drawImage in Java 1.1. If the image is transparent, you can provide a background color that will seep through. Otherwise, the optional parameter is ignored. Unfortunately, you cannot use drawImage to rotate the image. If you do want to rotate the image, you have to do more work and subclass ImageFilter, just like with Java 1.0.

8.7. ADDITIONAL PACKAGES

8.7.1. Java Utility Package

The final Java package, java.util, contains a collection of utility classes.

- **Data Structure Classes**

  A variety of useful classes implementing standard computer science data structures: including BitSet, Dictionary, Hashtable, Stack and Vector. The java.util package also defines the Enumeration interface which provides a protocol for classes to count through a set of values.

- **Date**

  Use the Date class to create and manipulate calendar dates in a system-independent fashion.

- **StringTokenizer**

  This class converts a String of text into its tokens.

- **Properties**

  This class implements persistent properties. The properties table contains key/value pairs where both the key and the value are Strings. This class is used by the System class to implement System properties.

- **Observer and Observable**

  Classes that implement the Observer interface can "watch" Observable objects for state changes. When an Observable object changes it notifies all of its Observers of the change.

- **Random-Number Generator**

  The Random class implements a random-number generator.

- **Enumeration**

  The Enumeration interface defines a generic programming interface for iterating through a set of values.

8.7.2. Java Language Package

By default, each Java application/applet has access to the *java.lang* package. Inside *java.lang* are classes that represent primitive data types (such as int & char), as well as more complex classes. It contains classes pertaining to strings, string buffers, threads, and even the System class from which we obtain out input and output streams. *Java.lang* is quite extensive, and some aspects (such as threads) can be confusing for those new to the Java language. For this reason, we'll only present the more useful, less complex, parts of this package. This should allow you to experiment with some of the handy features of the Java language, while minimising the barrier to understanding. For more advanced Java programmers, you may wish to consult the Java API yourself, and experiment with other parts of *java.lang* not covered herein.

Interfaces in java.lang

- Cloneable
- Runnable

Classes in java.lang

- Boolean
- Byte

- Character
- Class
- ClassLoader
- Compiler
- Double
- Float
- Integer
- Long
- Math
- Number
- Object
- Process
- Runtime
- SecurityManager
- Short
- String
- StringBuffer
- System
- Thread
- ThreadGroup
- Throwable
- Void

8.7.3. Java Collections Package

Collections are objects that hold other objects that are accessed, placed, and maintained under some set of rules. There are many different ways you could classify these different rules. This package uses one of the simplest, splitting them into four main categories, (Sets, Bags, Seqs, Maps), plus a few 'mixin' variants. Top-level functionality is described entirely via interfaces, *not* classes. This allows some mixing and matching of capabilities at the implementation level, and also enables you to add your own alternative implementations of this functionality that will interoperate with the versions provided in this package.

As a *user* of the collection package, you do not need to know a lot about data structures, but you do need to know what kinds of access and placement rules you want in a particular situation. The basic choices are:

- Sets contain at most one occurrence of any element.
- Bags may contain multiple occurrences of any element.
- Seqs are ordered, and have elements that are accessible via array-like indices.
- Maps associate unique keys with each element.
- l ElementSorted Collections maintain elements in order with respect to a given Comparator (sorting) function; KeySorted collections maintain them in order of their keys. (Any of the above kinds of collections may be element-sorted, but only Maps can be key-sorted.)

The base interface Collection reflects commonalities across all of these of kinds of rules. For reasons described below, all of the interfaces in the following diagram describe classes/objects that need not even be updatable (modifiable), so restrict themselves to operations that report properties of Collections or build new Collections with various properties.

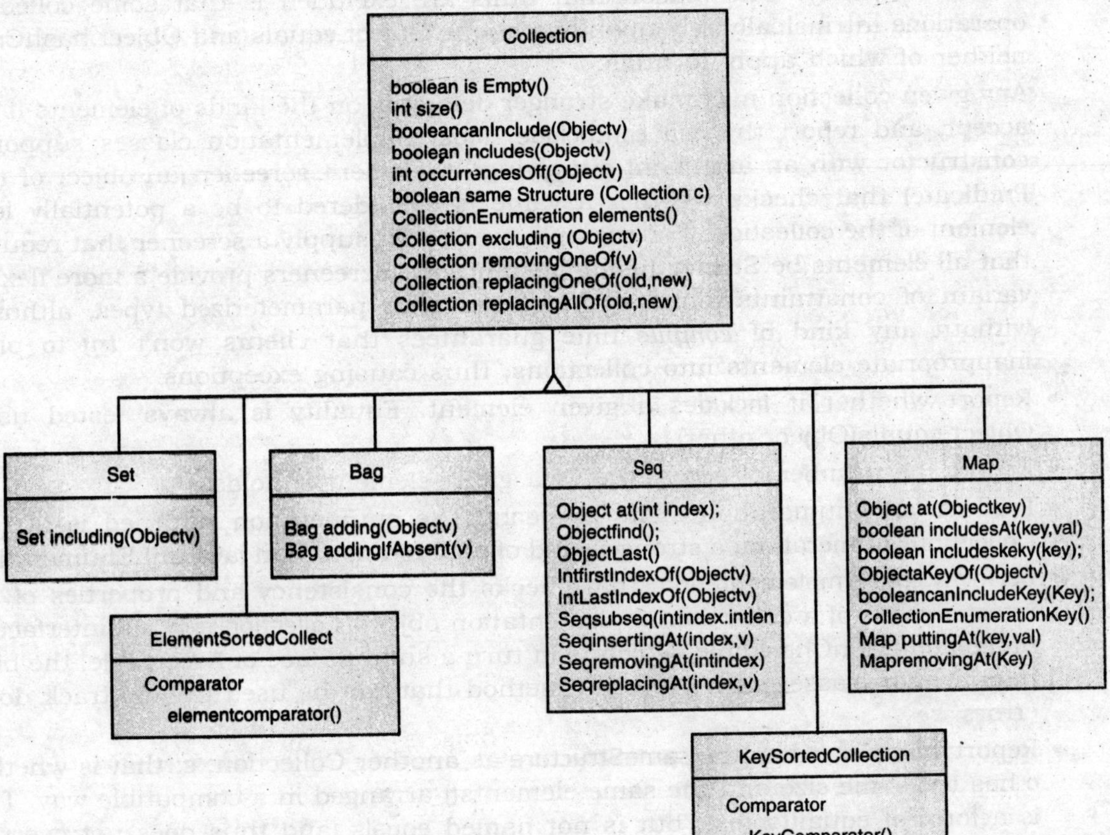

Fig. 8.27. Collection interface

Collection

Collection is the base interface for all collection classes defined in this package. Every collection can:

- Report its size and whether it isEmpty.
- Report whether it canInclude a given value.

  In general Collections hold Objects. Elements may be subclasses of Object, but not builtin scalars like int. You can, however, use numbers as elements via the java.lang.Number class (or rather its subclasses) to "wrap" the numbers as number-holding objects first. For example:

```
void addFloat(UpdatableBag b, float f)
{
    b.add(new Float(f));
}
```

Also, null is never a valid element, so canInclude(null) is always false. Operations screen for nulls, and raise exceptions upon attempts to add them to collections (but NOT when testing for their presence). This can be annoying, but is the most defensible policy. One reason that nulls are excluded is that some collection operations intrinsically rely upon operations Object.equals and Object.hashCode, neither of which apply to nulls.

Any given collection may make stronger demands on the kinds of elements it will accept, and report this via canInclude. Most implementation classes support a constructor with an argument supplying an element screener (an object of type Predicate) that checks whether a value is considered to be a potentially legal element of the collection. For example, you could supply a screener that requires that all elements be Strings (using 'instanceof'). Screeners provide a more flexible variant of constraints found in languages with parameterized types, although without any kind of *compile* time guarantees that clients won't *try* to place inappropriate elements into collections, thus causing exceptions.

- Report whether it includes a given element. Equality is always tested using Object.equals(Object other).
- Report the number of occurrencesOf a given element it holds.
- Produce an enumeration of its elements. The enumeration returned is of type CollectionEnumeration, a stronger kind of enumeration than java.util.Enumeration.
- Perform checkImplementation, that checks the consistency and properties of the internal state of a collection implementation object. Collection is a subinterface of ImplementationCheckable, which is in turn a subinterface of Assertable, the base interface for classes with an assert method that can be used to help track down errors.
- Report whether it has the sameStructure as another Collection, c; that is whether c has the same size and the same elements() arranged in a compatible way. This is a form of equality test, but is not named equals (and thus does not override Object.equals) since it is not always the sense of equality that you would want to apply to collections.
- Build a new Collection that is just like self except excluding (i.e., removing all occurrencesOf) a given element. This does *not* alter the Collection itself, it just makes a new one. This operation is a lot like *subtraction* of numbers. When you say a - b for two numbers a and b, you get a new number that represents the difference between them without changing a or b. The excluding operation has the same kind of effect. By convention, all such creational methods are named with gerunds (e.g., including, excluding, adding).

If you want a collection in which you actually change the contents of the current collection rather than building a new one, you'll need an UpdatableCollection with an exclude operation, described below. But even without using UpdatableCollections, you can still create *series* of collections, each one differing them the previous one by virtue of not including a given element via code like:

```
Collection excludingElements(Collection c, Enumeration e)
  {
      while (e.hasMoreElements())
```

```
        c = c.excluding(e.nextElement());
        return c;
    }
```

For many implementations of Collection, this is a ridiculously slow way to do it. However, as described below, others (the Incr classes) are tuned to do this reasonably efficiently.

- Build a new Collection that is just like self except removingOneOf (i.e., removing at most one occurrence of) a given element.
- Build a new Collection that is just like self except replacingOneOf (i.e., at most one occurrence of) some old value with a new value.
- Build a new Collection that is just like self except replacingAllOf (i.e., all occurrences of) some old value with a new value.
- Build a duplicate of self — an independent copy of the collection, but not necessarily its elements. Elements themselves are not necessarily cloned. (Note: duplicate is simply the externally callable version of method java.lang.Object.clone().)
- Support Serialization. All Collections, as well as other auxiliary classes and interfaces are declared to implement java.io.Serializable.
- Also, while it cannot be stated in the interface, each collection implementation supports:
  1. A default constructor, initializing it to an empty (and unscreened) collection;
  2. Synchronization. As a rule, all operations (even "read-only" operations) within implementation classes are declared as synchronized.

Sets

Sets are collections in which the number of occurrences of any object is at most one. They support an operation that causes includes to be true; i.e., adding an element only if it is not present.

Bags

Bags may contain multiple occurrences of each element. Bags can do anything that Sets can and more, except that they do not intrinsically guarantee that the number of occurrences of any element is at most one. (However, you can always use a Bag in a way that 'manually' enforces that the number of occurrences is at most one.)

Seqs

Seqs maintain elements in sequential order. They are sequential lists that can be accessed by positional indices ranging from 0..size()-1. Other Collections that are not Seqs may guarantee some kind of sequential ordering of elements, but don't support access of these elements via indices.

Maps

Maps are collections in which each element has a key. Any kind of Object may serve as a key for an element. (While you could use regular Collections in which each element holds a pair, Maps are generally simpler, more efficient, and contain operations useful under the interpretation that the pairs represent keyed elements.)

Maps also serve as a way to use a certain field of objects of some class as a key for purposes of storing in a collection. For example, supposing you want to use some accessor prop() defined in your class as the basis for entering and finding elements, you could use a Map, entering pairs of (x.prop(), x).

ElementSortedCollections

A mixing for collections that always maintain elements in sorted order with respect to a given Comparator Note: SortedCollections rely on the ordering properties of elements *NOT* to change once they have been placed in the collection. They cannot in general deal with cases where you change any of their elements behind their backs in ways that break their sort order. (See however, implementation checks that can help you discover if this occurred.)

ElementSorted Collections provide a different way to use properties of objects as their access keys. You can define your class to implement interface Keyed, with a method key() that returns the key (for example it might just return prop()). Then use a Comparator that compares on keys(). The DefaultComparator class does this automatically before trying to perform comparsions, but you can write your own Comparator that does this in a more specialized and efficient way.)

KeySortedCollections

A mixing for Maps that always maintain keys in sorted order

SUMMARY

Abstract window took kit is a portable GUI library.

- It can work on Solaris, Window 95/NT and Mac system 7.x. for stand alone applications and /or applets.
- We have explained more of the features of AWT/GUI hierarchy.
- We have also explained addition packages for common use.

SOLVED PROBLEMS

Q.8.1. Which containers use a border layout as their default layout?

Ans. The window, Frame and Dialog classes use a border layout as their default layout.

Q.8.2. What method is used to specify a container's layout?

Ans. The setLayout() method is used to specify a container's layout.

Q.8.3. Which containers use a FlowLayout as their default layout?

Ans. The Panel and Applet classes use the FlowLayout as their default layout.

Q.8.4. What's the difference between a GridLayout and a GridBagLayout?

Ans. A GridLayout places each component into a cell of one size. Thus all components have the same size. A GridbagLayout allows components to span multiple cells. Therefore components can have different sizes. There are other differences, but that's the main one.

Q.8.5. What's the difference between a Frame and a Dialog?

Ans. A Dialog is primarily for transitory and immediate responses. A Frame may last for a while.

Q.8.6. What's the difference between a Menu and a MenuBar?

Ans. A MenuBar contains menus.

Q.8.7. What's the difference between a Menu and a MenuItem?

Ans. Actually a Menu is a subclass of MenuItem. This allows you to make hierarchical menus by addinf menus to menus. However a MenuItem is generally thought of as a particular selection in a menu.

Q.8.8. Why wouldn't we create a Menu inside the MenuBar's add method like you create MenuItem's inside a Menu's add method?

Ans. If you did this you would not have a convenient reference to the Menu, so it would be difficult if not impossible to add items to the menu.This isn't an issue for MenuItems since they generate events, but Menus don't generate events.

EXERCISES

Q.8.1. What is the concept of AWT?

Q.8.2. Give the code to create following controls.

- (*a*) Button
- (*b*) Canvas
- (*c*) Check box
- (*d*) CheckboxGroup
- (*e*) Choice
- (*f*) Label
- (*g*) List
- (*h*) Scroll bar
- (*i*) Textfield
- (*j*) TextArea

Q.8.3. What are the difference between following

- (*a*) Check box and CheckboxGroup
- (*b*) Textfield and TextArea
- (*c*) GridLayout and GridBagLayout
- (*d*) Menu and MenuItem.

Q.8.4. Give the code to flip the image as follows:

- (*a*) flip an image alog the horizental axis
- (*b*) flip an image alog the vertical axis
- (*c*) flip an image alog the both horizental and vertical axis.

Q.8.5. Explain the following classes of utility package.

- (*a*) Hashtable
- (*b*) Vector
- (*c*) Random
- (*e*) Enumeration
- (*f*) Dictionary

❑❑❑

Chapter 9

Event Handling

9.1. INTRODUCTION

CHAPTER OUTLINE

- Introduction
- ActionListener Example
- Item Listener Example
- Adjustment Listener Example
- Key Listener and Key Adapter Example
- MouseMotion Listener
- Mouse Listener
- Window Listener
- FocusListener

Your actions when you're using the GUI for a window-based program or an applet-clicking a menu item or a button, moving the mouse, and so on-are first recognized by the operating system. For each action, the operating system determines which of the programs currently running on your computer should know about it and passes the action on to that program. When you click a mouse button, it's the operating system that registers this and notes the position of the mouse cursor on the screen. It then decides which application controls the window where the cursor was when you pressed the button, and communicates the mouse button-press to that program. The signals that a program receives from the operating system as a result of your actions are called events. The basic idea of how actions and events are communicated to your program code is illustrated in Figure 9.1.

Each event is some specific situation generated by keyboard, mouse or system itself.

For event handling there are some important concepts:

Event Source – that generate the events.

Event Listener – that listen the event.

Event handling – way to handling the events.

Java contains 11 Listener interfaces and 7 Adapter classes for event handling.

Sun Micro systems added the Adapter classes in the java JDK 1.2. Java provides Adapter classes for only those interface, which we have more then one method. Using Adapter classes user can override methods according to his/her need may be one/more then one or may be all.

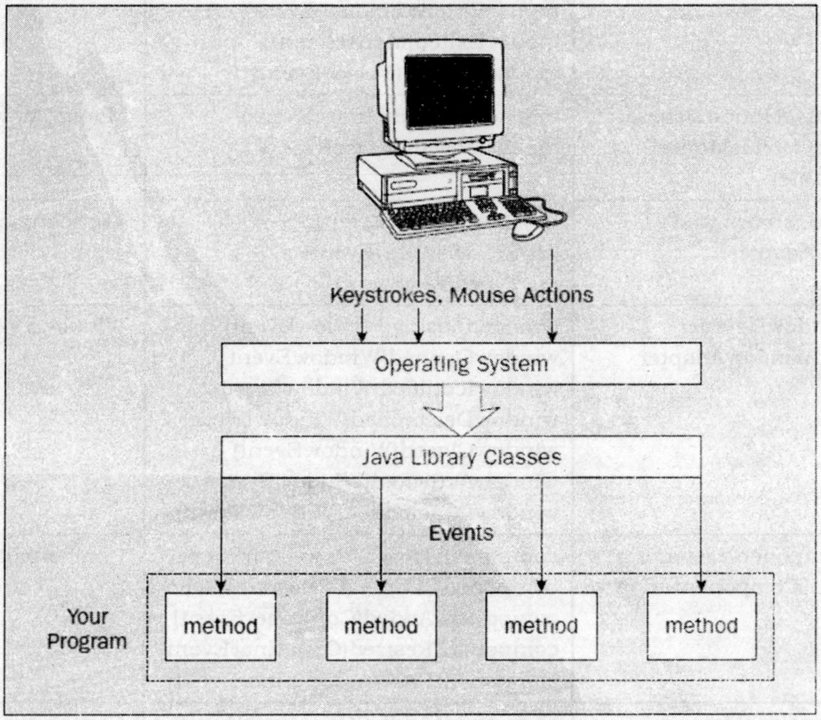

Fig. 9.1.

Concept

Any action performed by the user from input device, that is an event.

Here is the list of Listener interface and Adapter classes.

Category	Listeners / Adapters	Methods	Controls to used
Action	ActionListener	actionPerformed (ActionEvent)	Button Click / menu item / enter key / double click on any item
Item	ItemListener	itemStateChanged (ItemEvent)	Radio / List / Choice / CheckBox
Text	TextListener	textValueChanged (TesxtEvent)	when ever text value changed
Adjustment	AdjustmentListener	adjustmentValueChanged (AdjustmentEvent)	with Scrollbar

Focus	FocusListener and FocusAdapter	focusGained(FocusEvent) focusLost(FocusEvent)	Focus on any control
Mouse	MouseListener and MouseAdapter	mousePressed(MouseEvent) mouseReleased(MouseEvent) mouseEntered(MouseEvent) mouseExited(MouseEvent) mouseClicked(MouseEvent)	Mouse Clicks
Mouse Motion	MouseMotionListener and MouseMotion Adapter	mouseDragged(MouseEvent) mouseMoved(MouseEvent)	Mouse Movement
Key	KeyListener and KeyAdapter	keyPressed(KeyEvent) keyReleased(KeyEvent) keyTyped(KeyEvent)	Keyboard strikes
Window	WindowListener andWindowAdapter	windowClosing(WindowEvent) windowOpened(WindowEvent) windowIconified(WindowEvent) windowDeiconified(WindowEvent) windowClosed(WindowEvent) windowActivated(WindowEvent) windowDeactivated(WindowEvent)	Windows controls
Component	ComponentListener and ComponentAdapter	componentMoved(ConatinerEvent) componentHidden(ConatinerEvent) componentAdded(ConatinerEvent) componentResized(ConatinerEvent) componentShown(ConatinerEvent)	Components
Conatiner	ContainerListener andContainerAdapter	componentAdded(ConatinerEvent) componentRemoved(ConatinerEvent)	Containers

9.2. ACTIONLISTENER EXAMPLE

An ActionListener interface is basically used for Button, Menu items, Enter Key and the Double click of mouse controls. The ActionListener interface contains only one method for handling the event, which is :

```
public void actionPerformed(ActionEvent e){}
```

Now how to use this method in different programs, see some examples:

Example 9.1: Program to input two numbers using text boxes, create buttons for "+" ,"-", "*" , "/" and perform the action on them using event handling.

```
import java.lang.*;
import java.awt.*;
import java.awt.event.*;
import java.applet.*;

public class Example extends Applet implements ActionListener
```

```
{
Label l1;
TextField t1,t2,t3;
Button b1,b2,b3,b4;
public void init()
{
l1=new Label("Enter value ");
b1=new Button ("+");
b2=new Button("-");
b3=new Button("*");
b4=new Button("/");
t1=new TextField(10);
t2=new TextField(10);
t3=new TextField(10);
setBackground(new Color(200,100,200));

add(l1);

add(t1);
add(t2);
add(t3);
add(b1);
add(b2);
add(b3);
add(b4);
b1.addActionListener(this);
b2.addActionListener(this);

b3.addActionListener(this);
b4.addActionListener(this);
}
public void actionPerformed(ActionEvent e)
{
int a=Integer.parseInt(t1.getText());
int b=Integer.parseInt(t2.getText());
int c=0;
if(e.getSource()==b1)
{
    c=a+b;
}
if(e.getSource()==b2)
{
c=a-b;
}
if(e.getSource()==b3)
{
c=a*b;
}
```

```
if(e.getSource()==b4)
{
    c=a/b;
}
t3.setText(String.valueOf(c));
}
}
//<applet code=Example height=500 width=300></applet>
```

COMPILE:

```
C:\>javac Example.java
```

EXECUTE:

```
C:\>appletviewer Example.java
```

OUTPUT

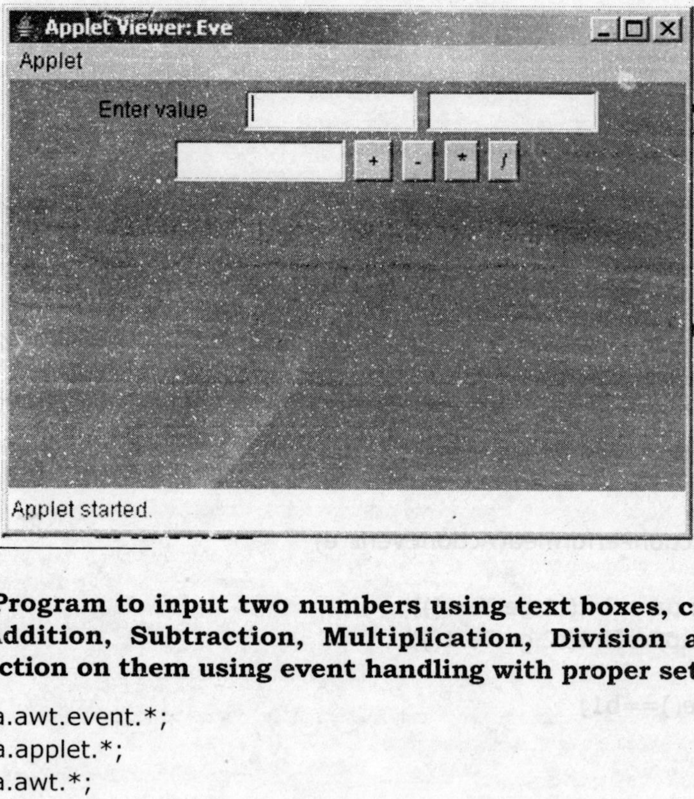

Example 9.2: Program to input two numbers using text boxes, create buttons for Addition, Subtraction, Multiplication, Division and perform the action on them using event handling with proper settings of controls.

```
import java.awt.event.*;
import java.applet.*;
import java.awt.*;

public class operation extends Applet implements ActionListener
{
    Label L1, L2,L3;
    TextField T1, T2, T3;
    Button b1,b2,b3,b4;
    public void init()
    {
```

```
        setLayout(null);
        L1=new Label("Enter first number:");
        L2=new Label("Enter second number:");
        L3=new Label("RESULT");

        T1=new TextField();
        T2=new TextField();
        T3=new TextField();

        b1=new Button("add");
        b2=new Button("Substract");
        b3=new Button("multiply");
        b4=new Button("divide");

        L1.setBounds(50,50,110,40);
        L2.setBounds(50,100,140,40);
        L3.setBounds(50,150,120,40);

        T1.setBounds(210,50,50,30);
        T2.setBounds(210,100,50,30);
        T3.setBounds(210,150,50,30);
        b1.setBounds(50,250,50,30);
        b2.setBounds(120,250,60,30);
        b3.setBounds(190,250,50,30);
        b4.setBounds(260,250,50,30);
        add (L1);
        add (L2);
        add (L3);
        add(T1);
        add(T2);
        add(T3);
        add(b1);
        add(b2);
        add(b3);
        add(b4);
        b1.addActionListener(this);
        b2.addActionListener(this);
        b3.addActionListener(this);
        b4.addActionListener(this);
    }
    public void actionPerformed(ActionEvent e)
    {
        int a=Integer.parseInt(T1.getText());
        int b=Integer.parseInt(T2.getText());
        int c=0;
        if(e.getSource()==b1)
            c =a + b;
        if(e.getSource()==b2)
```

```
            c=a-b;
        if(e.getSource()==b3) c=a*b;
        if(e.getSource()==b4)c=a/b;
        T3.setText(String.valueOf(c));
    }
}
//<applet code = operation.class width=500 height=500></applet>
```

COMPILE:

```
C:\>javac operation.java
```

EXECUTE:

```
C:\>appletviewer operation.java
```

OUTPUT

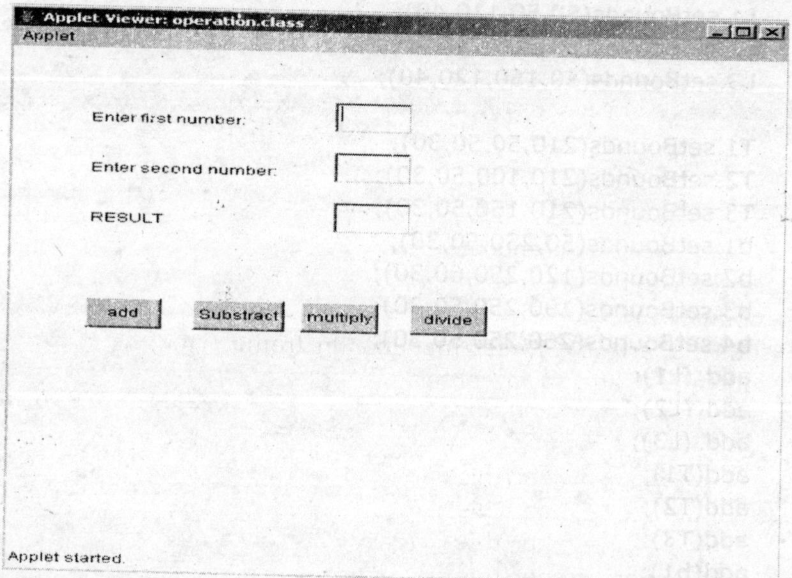

Example 9.3: An AWT Program to create a Note-Pad with Menu items and perform the action on them using event handling.

```
import java.awt.*;
import java.awt.event.*; // step - 1
class Notepad extends Frame implements ActionListener  // step - 2
{
    FileDialog fd; // for open & save Dialog
    TextArea ta = new TextArea();
    MenuBar mb = new MenuBar();
    Menu f1 = new Menu("F i l e ");
    Menu f2 = new Menu("E d i t ");
    MenuItem a1 = new MenuItem("N e w ");
    MenuItem a2 = new MenuItem("O p e n ");
    MenuItem a3 = new MenuItem("S a v e ....");
```

```
MenuItem a4 = new MenuItem("E x i t ");
MenuItem a5 = new MenuItem("C u t ");
MenuItem a6 = new MenuItem("C o p y ");
MenuItem a7 = new MenuItem("P a s t e ");
MenuItem a8 = new MenuItem("S e l e c t  A l l ");
Notepad()
{
    setTitle("Notepad Example By Lalit Arora");
    setFont(new Font("Courier New",1,14));
    // add item in file menu
    f1.add(a1);
    f1.add(a2);
    f1.add(a3);
    f1.addSeparator();
    f1.add(a4);
    // add item in edit menu
    f2.add(a5);
    f2.add(a6);
    f2.add(a7);
    f2.addSeparator();
    f2.add(a8);
    // add menu in menu bar
    mb.add(f1);
    mb.add(f2);
    setMenuBar(mb); // set menubar in frame
    add(ta); // add ta in center
    setBounds(50,50,700,500); // set size of notepad
    setVisible(true);
    a1.addActionListener(this);
    a2.addActionListener(this);
    a3.addActionListener(this);
    a4.addActionListener(this);
    a5.addActionListener(this);
    a6.addActionListener(this);
    a7.addActionListener(this);
    a8.addActionListener(this);
    addWindowListener(new Abhi());
}
class Abhi extends WindowAdapter
{
    public void windowClosing(WindowEvent e)
    {
        dispose();
    }
}
public void actionPerformed(ActionEvent e)
{
    if(e.getSource()==a1) // for new
```

```
        {
            setTitle("Untited");
            ta.setText("");
        }
        if(e.getSource()==a2) // for open
        {
            fd = new FileDialog(this,"Open Me",FileDialog.LOAD);
            fd.setVisible(true);
            setTitle(fd.getFile());
        }
        if(e.getSource()==a3) // for save
        {
            fd = new FileDialog(this,"Save Me",FileDialog.SAVE);
            fd.setVisible(true);
            setTitle(fd.getFile());
        }
        if(e.getSource()==a4) // for exit
            dispose(); // setVisible(false);
        // mar gaya  / jinda hai
    }
    public static void main(String aa[])
    {
        new Notepad();
    }
}
```

COMPILE:

C:\>javac Example.java

EXECUTE:

C:\>java Example

OUTPUT

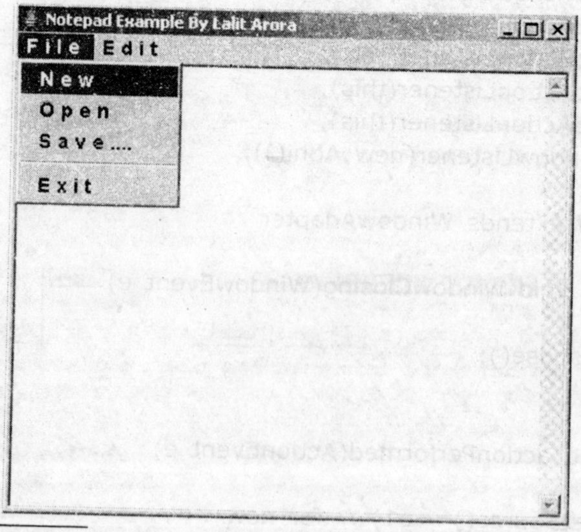

Example 9.4 Program to play Sounds files on applet.

```
import java.awt.*;
import java.awt.event.*;
import java.applet.*;
public class Sounds extends Applet implements ActionListener
{
    AudioClip c1,c2,c3;
    Button a1 = new Button("Hi !");
    Button a2 = new Button("Yahoo");
    Button a3 = new Button("Music");
    public void init()
    {
        c1 = getAudioClip(getCodeBase(),"hi.au");
        c2 = getAudioClip(getCodeBase(),"yahoo1.au");
        c3 = getAudioClip(getCodeBase(),"spacemusic.au");
        add(a1);
        add(a2);
        add(a3);
        a1.addActionListener(this);
        a2.addActionListener(this);
        a3.addActionListener(this);
    }
    public void actionPerformed(ActionEvent e)
    {
        if(e.getSource()==a1)
            c1.play(); // loop(); / stop();
        if(e.getSource()==a2)
            c2.play();
        if(e.getSource()==a3)
            c3.play();
    }
}
//<applet code=Sounds height=400 width=600></applet>
```

COMPILE:

```
C:\>javac Sounds.java
```

EXECUTE:

```
C:\>appletviewer Sounds.java
```

OUTPUT:

> **Note:** Here we used the sounds files "hi.au","yahoo1.au", "spacemusic.au" existing in the same directory of program. You can use your own file (*.au - audio files) or you can download from internet.

9.3. ITEMLISTENER EXAMPLE

An ItemListener interface is basically used for Radio, List, Choice and the Check Box controls. The ItemListener interface contains only one method for handling the event, which is :

```
public public void itemStateChanged(ItemEvent e){}
```

Now how to use this method in different programs, see some examples:

Example 9.5: Program to create Radio buttons to select gender and perform the action on them using event handling and display gender on status-bar.

```
import java.awt.*;
import java.awt.event.*;
import java.applet.*;
public class Radio extends Applet implements ItemListener
{
    CheckboxGroup cg = new CheckboxGroup(); // for grouping
    Checkbox c1 = new Checkbox("Male.",cg,true); // label ,group,state
    Checkbox c2 = new Checkbox("Female.",cg,false);
    public void init()
    {
        setFont(new Font("Courier New",1,14));
        setBackground(new Color(170,194,250));//Color.pink
        add(c1);
        add(c2);
        c1.addItemListener(this);
        c2.addItemListener(this);
    }
    public void itemStateChanged(ItemEvent e)
    {
        if(c1.getState()) // true if checked else return false
            showStatus("U r select - "+c1.getLabel());
        if(c2.getState()) // true if checked else return false
            showStatus("U r select - "+c2.getLabel());
    }
}
//<applet code=Radio height=400 width=600></applet>
```

COMPILE:

```
C:\>javac Radio.java
```

EXECUTE:

```
C:\>appletviewer Radio.java
```

OUTPUT

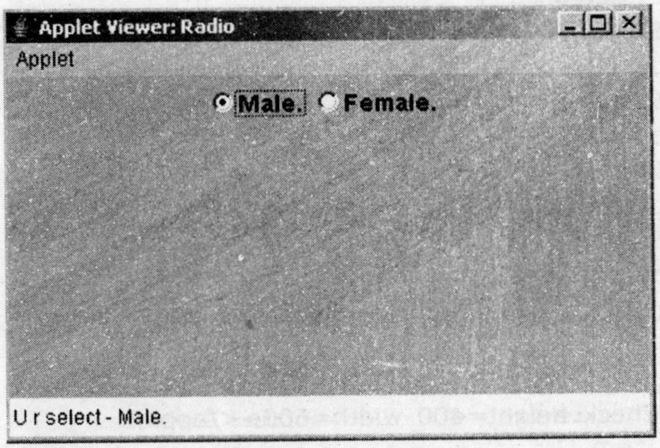

Example 9.6: Program to create Check Boxes for qualification and display the selected qualifications in Textbox.

```
import java.awt.*;
import java.awt.event.*;
import java.applet.*;
public class Check extends Applet implements ItemListener
{
    Checkbox c1 = new Checkbox("BSc.",true); // true for checked, default is false
    Checkbox c2 = new Checkbox("MSc.");
    Checkbox c3 = new Checkbox("B.Tech.");
    Checkbox c4 = new Checkbox("MCA.");
    Checkbox c5 = new Checkbox("MBA.");
    TextField t1 = new TextField(50); // for 50 chars
    public void init()
    {
        setFont(new Font("Courier New",1,14));
        setBackground(new Color(170,194,250));
        add(c1);
        add(c2);
        add(c3);
        add(c4);
        add(c5);
        add(t1);
        c1.addItemListener(this);
        c2.addItemListener(this);
        c3.addItemListener(this);
        c4.addItemListener(this);
        c5.addItemListener(this);
    }
    public void itemStateChanged(ItemEvent e)
    {
        String s = "Qualification -> ";
```

```
            if(c1.getState())
                s+=c1.getLabel()+" , ";
            if(c2.getState()) // true if checked else return false
                s+=c2.getLabel()+" , ";
            if(c3.getState())
                s+=c3.getLabel()+" , ";
            if(c4.getState())
                s+=c4.getLabel()+" , ";
            if(c5.getState())
                s+=c5.getLabel();
            t1.setText(s);
        }
    }
    //<applet code=Check height=400 width=600></applet>
```

COMPILE:

```
C:\>javac Check.java
```

EXECUTE:

```
C:\>appletviewer Check.java
```

OUTPUT

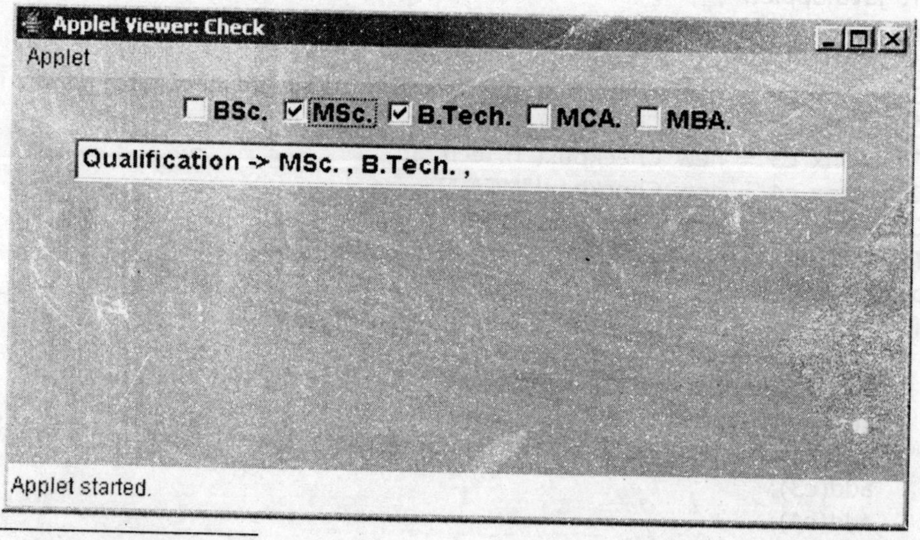

Example 9.7: Program to create List Box and Combo Box and display the selected items in Text Box.

```
import java.awt.*;
import java.awt.event.*;
import java.applet.*;
public class MyList extends Applet implements ItemListener
{
    List a = new List(3); // no of rows visible , default is 4
    Choice c = new Choice();
```

```
        TextField t1 = new TextField(50);
        String ar[] = new String[]{"Noida","Ghaziabad","Delhi","New
Delhi","Mathura","Gorakhpur","Lakhimpur","Basti","Nagpur"};
        public void init()
        {
            setFont(new Font("Courier New",1,14));
            setBackground(new Color(170,194,250));
            for(int i=0;i<ar.length;i++)
                a.add(ar[i]);
            for(int i=1970;i<=2008;i++)
                c.add(""+i);
            add(a);
            add(c);
            add(t1);
            c.select("1986");
            c.addItemListener(this);
            a.addItemListener(this);
        }
        public void itemStateChanged(ItemEvent e)
        {
            t1.setText("U r from - "+a.getSelectedItem()+" , selected year is - "
+c.getSelectedItem());
        }
    }
    //<applet code=MyList height=400 width=600></applet>
```

COMPILE:

```
C:\>javac MyList.java
```

EXECUTE:

```
C:\>appletviewer MyList.java
```

OUTPUT:

9.4. ADJUSTMENTLISTENER EXAMPLE

An AdjustmentListener interface is basically used only for Scrollbar control. The AdjustmentListener interface contains only one method for handling the event, which is :

```
public void adjustmentValueChanged(AdjusmentEvent e){}
```

Now how to use this method in different programs, see some examples:

Example 9.8: Program to create three Scrollbars for Red, Green, Blue Colors and change the color of back ground according to adjustment of scrollbars.

```
import java.awt.*;
import java.awt.event.*;
import java.applet.*;
public class Scroll extends Applet implements AdjustmentListener
{   //type(0-H/1-V),initial,thumb,min,max(scrollmax=max-thumb)
    Scrollbar r = new Scrollbar(0,0,20,0,275);
    Scrollbar g = new Scrollbar(0,0,20,0,275);
    Scrollbar b = new Scrollbar(0,0,20,0,275);
    int cr,cg,cb;
    public void init()
    {
        setLayout(null);
        r.setBounds(50,50,600,20);
        g.setBounds(50,75,600,20);
        b.setBounds(50,100,600,20);
        r.setBackground(Color.red);
        g.setBackground(Color.green);
        b.setBackground(Color.blue);
        add(r);
        add(g);
        add(b);
        r.addAdjustmentListener(this);
        g.addAdjustmentListener(this);
        b.addAdjustmentListener(this);
    }
    public void adjustmentValueChanged(AdjustmentEvent e)
    {
        cr = r.getValue();
        cg = g.getValue();
        cb = b.getValue();
        setBackground(new Color(cr,cg,cb));
        showStatus("Color is --> r = "+cr+" , g = "+cg+" , b = "+cb);
    }
}
//<applet code=Scroll height=400 width=600></applet>
```

COMPILE:

```
C:\>javac Scroll.java
```

EXECUTE:

```
C:\>appletviewer Scroll.java
```

OUTPUT

9.5. KEYLISTENER AND KEYADAPTER EXAMPLE

An KeyListener interface and KeyAdapter class is basically used only for keys of key board. The KeyListener interface and KeyAdapter class contains following methods for handling the event, which are :

```
public void keyPressed(KeyEvent e){}
public void keyTyped(KeyEvent e){}
public void keyReleased(KeyEvent e){}
```

Now how to use these methods in different programs, see some examples:

Example 9.9: Program to show the code of keys with their characters after pressing the individual key.

```
import java.awt.*;
import java.awt.event.*;
import java.applet.*;
import java.util.*;
public class Key extends Applet implements KeyListener
{
    int x;
    char c;
    public void init()
    {
        addKeyListener(this);
    }
    public void keyTyped(KeyEvent e){}
    public void keyReleased(KeyEvent e){}
    public void keyPressed(KeyEvent e)
    {
        x = e.getKeyCode(); // return key code not key ascii
        c = e.getKeyChar(); // return printable key char
        showStatus("Key Pressed , key code is --> "+x+" , key char is --> "+c);
    }
}
//<applet code=Key height=400 width=600></applet>
```

COMPILE:

```
C:\>javac Key.java
```

EXECUTE:

```
C:\>appletviewer Key.java
```

OUTPUT: (press any Key on the applet)

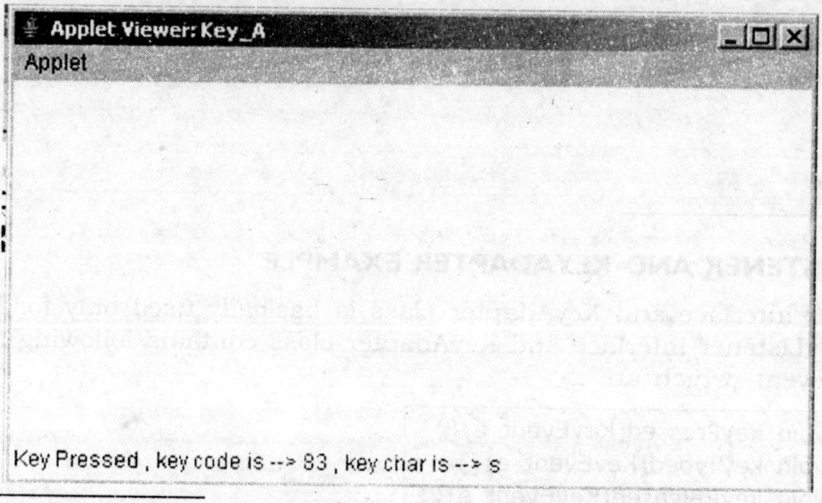

Example 9.10: Program to show the code of keys with their characters after pressing the individual key.

```
import java.awt.*;
import java.awt.event.*;
import java.applet.*;
public class Key_A extends Applet
{
    int x;
    char c;
    public void init()
    {
        addKeyListener(new Deepu());
    }
    class Deepu extends KeyAdapter
    {
        public void keyPressed(KeyEvent e)
        {
            x = e.getKeyCode(); // return key code not key ascii
            c = e.getKeyChar(); // return printable key char
            showStatus("Key Pressed , key code is --> "+x+" , key char is --> "+c);
        }
    }
}
//<applet code=Key_A height=400 width=600></applet>
```

COMPILE:

```
C:\>javac Key_A.java
```

EXECUTE:

```
C:\>appletviewer Key_A.java
```

OUTPUT: (press any Key on the applet)

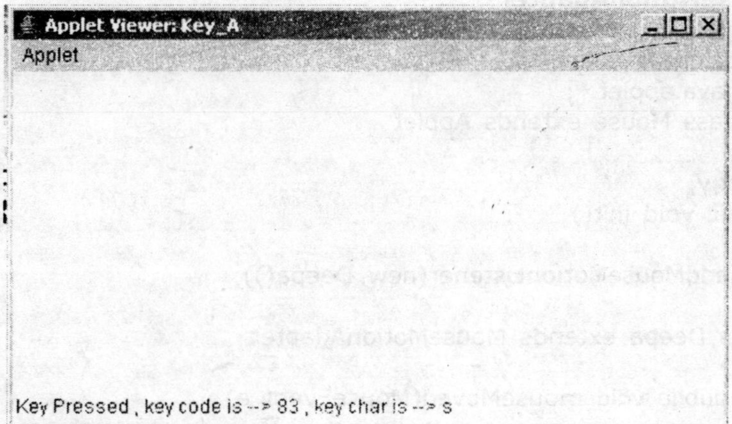

9.6. MOUSEMOTIONLISTENER

An MouseMotionListener interface and MouseMotionAdapter class is basically used for mouse movement. The MouseMotionListener interface and MouseMotionAdapter class contains only two methods for handling the event, which are :

```
public void mouseMoved(MouseEvent e){}
public void mouseDragged(MouseEvent e){}
```

Now how to use these methods in different programs, see some examples:

Example 9.11: Program to show the current status of Mouse Motion where mouse pointer will move around the applet.

Using MouseMotionListener

```
import java.awt.*;
import java.awt.event.*;
import java.applet.*;
public class Mouse extends Applet implements MouseMotionListener
{
    int x,y;
    public void init()
    {
        addMouseMotionListener(this);
    }
    public void mouseMoved(MouseEvent e)
    {
        x = e.getX(); // x hold the current mouse x position
        y = e.getY(); // y hold the current mouse y position
```

```
            showStatus("mouse Moved at --> x = "+x+" , y = "+y);
        }
        public void mouseDragged(MouseEvent e) { }
    }
    //<applet code=Mouse height=400 width=600></applet>
```

Using MouseMotionAdapter

```
    import java.awt.*;
    import java.awt.event.*;
    import java.applet.*;
    public class Mouse extends Applet
    {
        int x,y;
        public void init()
        {
            addMouseMotionListener(new Deepa());
        }
        class Deepa extends MouseMotionAdapter
        {
            public void mouseMoved(MouseEvent e)
            {
                x = e.getX(); // x hold the current mouse x position
                y = e.getY(); // y hold the current mouse y position
                showStatus("mouse Moved at --> x = "+x+" , y = "+y);
            }
        }
    }
    //<applet code=Mouse height=400 width=600></applet>
```

COMPILE:

```
C:\>javac Mouse.java
```

EXECUTE:

```
C:\>appletviewer Mouse.java
```

OUTPUT: (Move the mouse pointer on the applet)

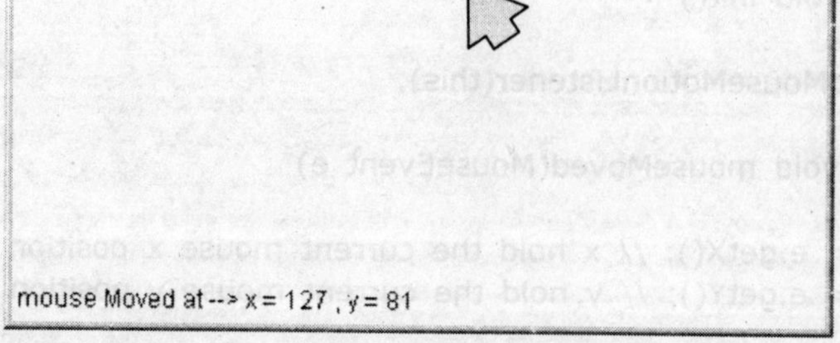

Example 9.12: Program to design a Simple Paint Brush on an applet.

```
import java.awt.*;
import java.awt.event.*;
import java.applet.*;
public class Mouse extends Applet implements MouseMotionListener
{
    Graphics g;
    int x,y;
    public void init()
    {
        g = getGraphics(); // g hold the current Applet graphics , now we can drawing
without using paint method
        addMouseMotionListener(this);
    }
    public void mouseMoved(MouseEvent e)
    {
        x = e.getX(); // x hold the current mouse x position
        y = e.getY(); // y hold the current mouse y position
        showStatus("mouse Moved at --> x = "+x+" , y = "+y);
    }
    public void mouseDragged(MouseEvent e)
    {
        x = e.getX(); // x hold the current mouse x position
        y = e.getY(); // y hold the current mouse y position
        showStatus("mouse dragged at --> x = "+x+" , y = "+y);
        g.fillOval(x,y,10,10);
        //g.drawString("Arora",x,y);
    }
}
//<applet code=Mouse height=400 width=600></applet>
```

COMPILE:

```
C:\>javac Mouse.java
```

EXECUTE:

```
C:\>appletviewer Mouse.java
```

OUTPUT: (Drag the mouse pointer on the applet to draw any image)

9.7. MOUSELISTENER

An MouseListener interface and MouseAdapter class is basically used for mouse movement. The MouseListener interface and MouseAdapter class contains following methods for handling the event, which are :

```
public void mousePressed(MouseEvent e){}
public void mouseClicked(MouseEvent e){}
public void mouseReleased(MouseEvent e){}
public void mouseEntered(MouseEvent e){} - over
public void mouseExited(MouseEvent e){} - out
```

Now how to use these methods in different programs, see some examples:

Example 9.13: Program to show all the methods of MouseListener on an applet.

```
import java.awt.*;
import java.awt.event.*;
import java.applet.*;
public class Mouse extends Applet implements MouseListener
{
    int cg,cb,cr;
    public void init()
    {
        addMouseListener(this);
    }
    public void mousePressed(MouseEvent e)
    {
        showStatus("mouse Pressed");
    }
```

```
        public void mouseReleased(MouseEvent e)
        {
            showStatus("mouse Released");
        }
        public void mouseEntered(MouseEvent e)
        {
            showStatus("mouse over");
        }
        public void mouseExited(MouseEvent e)
        {
            showStatus("mouse out");
        }
        public void mouseClicked(MouseEvent e)
        {
            showStatus("mouse Clicked");
        }
    }
    //<applet code=Mouse height=400 width=600></applet>
```

COMPILE:

```
C:\>javac Mouse.java
```

EXECUTE:

```
C:\>appletviewer Mouse.java
```

OUTPUT: (Move the mouse pointer on the applet)

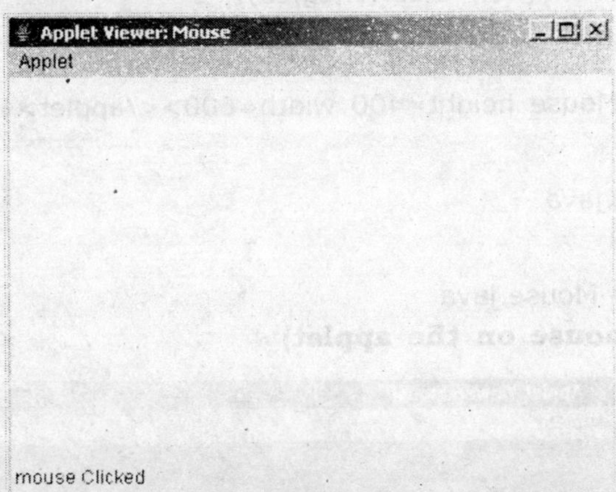

Example 9.14: Program to Change the back ground on Mouse click on an applet.

```
import java.awt.*;
import java.awt.event.*;
import java.applet.*;
public class Mouse extends Applet implements MouseListener
{
    int cg,cb,cr;
    public void init()
```

```
        {
            addMouseListener(this);
        }
        public void mousePressed(MouseEvent e)
        {
            showStatus("mouse Pressed");
        }
        public void mouseReleased(MouseEvent e)
        {
            showStatus("mouse Released");
        }
        public void mouseEntered(MouseEvent e)
        {
            showStatus("mouse over");
        }
        public void mouseExited(MouseEvent e)
        {
            showStatus("mouse out");
        }
        public void mouseClicked(MouseEvent e)
        {
            cr = (int)(Math.random()*255);
            cg = (int)(Math.random()*255);
            cb = (int)(Math.random()*255);
            setBackground(new Color(cr,cg,cb));
        }
    }
    //<applet code=Mouse height=400 width=600></applet>
```

COMPILE:

```
C:\>javac Mouse.java
```

EXECUTE:

```
C:\>appletviewer Mouse.java
```

OUTPUT: (Click the mouse on the applet)

9.8. WINDOWLISTENER

An WindowListener interface and WindowAdapter class is basically used for windows operations movement. The WindowListener interface and WindowAdapter class contains following methods for handling the event, which are :

```
public void windowOpened(WindowEvent e){}
public void windowClosing(WindowEvent e){}
public void windowClosed(WindowEvent e){}
public void windowActivated(WindowEvent e){}
public void windowDeactivated(WindowEvent e){}
public void windowIconified(WindowEvent e){} - minimize
public void windowDeiconified(WindowEvent e){} - maximize
```

Now how to use these methods in different programs, see some examples:

Example 9.15: An AWT Program to create a Note-Pad with Menu items and perform the action on them using event handling.

```
import java.awt.*;
import java.awt.event.*;
class Notepad extends Frame implements ActionListener
{
    FileDialog fd; // for open & save Dialog
    TextArea ta = new TextArea();
    MenuBar mb = new MenuBar();
    Menu f1 = new Menu("F i l e ");
    Menu f2 = new Menu("E d i t ");
    MenuItem a1 = new MenuItem("N e w ");
    MenuItem a2 = new MenuItem("O p e n ");
    MenuItem a3 = new MenuItem("S a v e ....");
    MenuItem a4 = new MenuItem("E x i t ");
    MenuItem a5 = new MenuItem("C u t ");
    MenuItem a6 = new MenuItem("C o p y ");
    MenuItem a7 = new MenuItem("P a s t e ");
    MenuItem a8 = new MenuItem("S e l e c t  A l l ");
    Notepad()
    {
        setTitle("Notepad Example By Lalit Arora");
        setFont(new Font("Courier New",1,14));
        // add item in file menu
        f1.add(a1);
        f1.add(a2);
        f1.add(a3);
        f1.addSeparator();
        f1.add(a4);
        // add item in edit menu
        f2.add(a5);
        f2.add(a6);
        f2.add(a7);
```

```
        f2.addSeparator();
        f2.add(a8);
        // add menu in menu bar
        mb.add(f1);
        mb.add(f2);
        setMenuBar(mb); // set menubar in frame
        add(ta); // add ta in center
        setBounds(50,50,700,500); // set size of notepad
        setVisible(true);
        a1.addActionListener(this);
        a2.addActionListener(this);
        a3.addActionListener(this);
        a4.addActionListener(this);
        a5.addActionListener(this);
        a6.addActionListener(this);
        a7.addActionListener(this);
        a8.addActionListener(this);
        addWindowListener(new Abhi());
    }
    class Abhi extends WindowAdapter
    {
        public void windowClosing(WindowEvent e)
        {
            dispose();
        }
    }
    public void actionPerformed(ActionEvent e)
    {
        if(e.getSource()==a1) // for new
        {
            setTitle("Untited");
            ta.setText("");
        }
        if(e.getSource()==a2) // for open
        {
            fd = new FileDialog(this,"Open Me",FileDialog.LOAD);
            fd.setVisible(true);
            setTitle(fd.getFile());
        }
        if(e.getSource()==a3) // for save
        {
            fd = new FileDialog(this,"Save Me",FileDialog.SAVE);
            fd.setVisible(true);
            setTitle(fd.getFile());
        }
        if(e.getSource()==a4) // for exit
            dispose(); // setVisible(false);
        // mar gaya  / jinda hai
```

```
    }
    public static void main(String aa[])
    {
        new Notepad();
    }
}
```

COMPILE:

```
C:\>javac Example.java
```

EXECUTE:

```
C:\>java Example
```

OUTPUT:

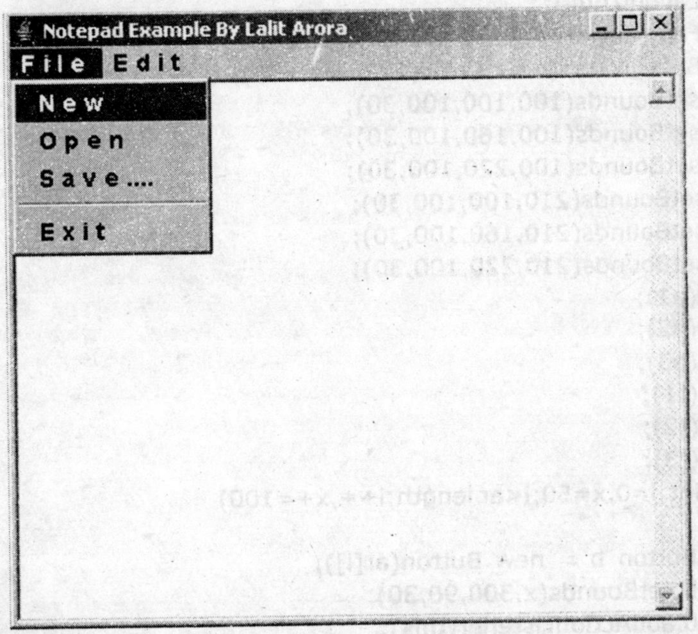

9.9. FOCUSLISTENER

An FocusListener interface and FocusAdapter class is basically used for Focus on Text box and Button controls. The FocusListener interface and FocusAdapter class contains only two methods for handling the event, which are :

```
public void focusGained(FocusEvent e){}
public void focusLost(FocusEvent e){}
```

Now how to use these methods in different programs, see some examples:

Example 9.16: Program to Add, Multiply, Subtract and Divide two numbers on applet.

```
import java.awt.*;
import java.awt.event.*;
```

```
import java.applet.*;
public class Focus extends Applet implements ActionListener,FocusListener
{
    String ar[] = new String[]{"Sum","Sub","Multi","Div","Clear"};
    Label a1 = new Label("Ist Num");
    Label a2 = new Label("IInd Num");
    Label a3 = new Label("Result");
    TextField t1 = new TextField();
    TextField t2 = new TextField();
    TextField t3 = new TextField();

    public void init()
    {
        setLayout(null);
        setFont(new Font("Courier New",1,14));
        setBackground(new Color(129,164,250));
        a1.setBounds(100,100,100,30);
        a2.setBounds(100,160,100,30);
        a3.setBounds(100,220,100,30);
        t1.setBounds(210,100,100,30);
        t2.setBounds(210,160,100,30);
        t3.setBounds(210,220,100,30);
        add(a1);
        add(a2);
        add(a3);
        add(t1);
        add(t2);
        add(t3);
        for(int i=0,x=50;i<ar.length;i++,x+=100)
        {
            Button b = new Button(ar[i]);
            b.setBounds(x,300,90,30);
            b.addActionListener(this);
            add(b);
        }
        t3.setEditable(false); // user can't edit now
        t1.addFocusListener(this);
        t2.addFocusListener(this);
    }
    public void focusGained(FocusEvent e)
    {
        if(e.getSource()==t1) // if focus gained on t1
        {
            t1.setText("");
            t2.setText("");
            t3.setText("");
        }
    }
```

```
public void focusLost(FocusEvent e)
{
    String s,n="";
    if(e.getSource()==t1) // if focus lost from t1
    {
        s = t1.getText();
        for(int i=0;i<s.length();i++)
        {
            if(s.charAt(i) != ' ')
                n+=s.charAt(i);
        }
        if(n.equals("")) t1.setText("0");
        else t1.setText(n);
    }
    if(e.getSource()==t2) // if focus lost from t2
    {
        s = t2.getText();
        for(int i=0;i<s.length();i++)
        {
            if(s.charAt(i) != ' ')
                n+=s.charAt(i);
        }
        if(n.equals("")) t2.setText("0");
        else t2.setText(n);
    }
}
public void actionPerformed(ActionEvent e)
{
    int a,b,c;
    String cmd = e.getActionCommand(); // return the button caption
    try{ a = Integer.parseInt(t1.getText()); }catch(Exception ex){a = 0;}
    try{ b = Integer.parseInt(t2.getText()); }catch(Exception ex){b = 0;}
    if(cmd.equals("Sum"))
    {
        c = a + b;
        t3.setText(""+c);
    }
    if(cmd.equals("Sub"))
    {
        c = a - b;
            t3.setText(""+c);
    }
    if(cmd.equals("Multi"))
    {
        c = a * b;
        t3.setText(""+c);
    }
    if(cmd.equals("Div"))
```

```
        {
            try
            {
                c = a / b;
                t3.setText(""+c);
            }catch(Exception ex){ t3.setText("Infinity");}
        }
        if(cmd.equals("Clear"))
        {
            t1.setText("");
            t2.setText("");
            t3.setText("");
        }
    }
}
//<applet code=Focus height=400 width=600></applet> COMPILE:
```

COMPILE:

```
C:\>javac Focus.java
```

EXECUTE:

```
C:\>appletviewer Focus.java
```

OUTPUT:

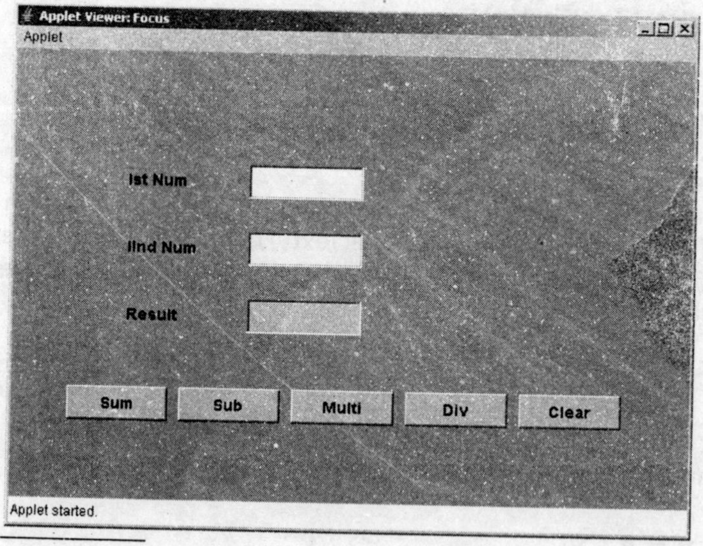

Example 9.17: Program of Employee Management System on applet.

```
import java.awt.event.*;
import java.awt.*;
import java.applet.*;
public class Focus extends Applet implements ActionListener,FocusListener
{
    String no[] = new String[50];
```

```
        String nm[] = new String[50];
        String sal[] = new String[50];
        int ctr=0,nor=0;
    String ar[] = new
String[]{"Exit","Cancel","Report","First","Prev","Next","Last","Delete","Update"};
        Label a1 = new Label("Empno");
        Label a2 = new Label("Ename");
        Label a3 = new Label("Salary");
        Label L  = new Label("System Message:");
        TextField t1 = new TextField();
        TextField t2 = new TextField();
        TextField t3 = new TextField();
        Button b1 = new Button("Save");
        public void init()
        {
            setLayout(null);
            setFont(new Font("Courier New",1,14));
            setBackground(new Color(129,164,250));
            a1.setBounds(100,100,100,30);
            a2.setBounds(100,160,100,30);
            a3.setBounds(100,220,100,30);
            L.setBounds(100,360,500,30);
            t1.setBounds(210,100,100,30);
            t2.setBounds(210,160,150,30);
            t3.setBounds(210,220,100,30);
            b1.setBounds(50,300,90,30);
            add(a1);
            add(a2);
            add(a3);
            add(L);
            add(t1);
            add(t2);
            add(t3);
            add(b1);
            for(int i=0,x=150;i<3;i++,x+=100)
            {
                Button b = new Button(ar[i]);
                b.setBounds(x,300,90,30);
                b.addActionListener(this);
                add(b);
            }
            for(int i=3,x=50;i<ar.length;i++,x+=50)
            {
                Button b = new Button(ar[i]);
                b.setBounds(450,x,90,30);
                b.addActionListener(this);
                add(b);
            }
```

```
        t1.addFocusListener(this);
        b1.addActionListener(this);
        b1.setEnabled(false);
    //  requestFocus(); // focus on frame
        }
        public void focusGained(FocusEvent e)
        {
            clear();
            b1.setEnabled(true);
        }
        public void focusLost(FocusEvent e)
        {
            if(t1.getText().equals(""))
                b1.setEnabled(false);
            for(int i=0;i<nor;i++)
            {
                if(t1.getText().equals(no[i]))
                {
                    disp(i);
            L.setText("System Message :Sorry U can't insert duplicate record , record
already exist");
            b1.setEnabled(false);
            break;
            }
        }
    }
    void save()
    {
        no[nor] = t1.getText();
        nm[nor] = t2.getText();
        sal[nor] = t3.getText();
        nor++; // ready for next record index.
    }
    void clear()
    {
        t1.setText("");
        t2.setText("");
        t3.setText("");
    }
    void disp(int a)
    {
        t1.setText(no[a]);
        t2.setText(nm[a]);
        t3.setText(sal[a]);
    }
    void updRec(int a)
    {
        nm[a] = t2.getText();
```

```
        sal[a] = t3.getText();
    }
    public void actionPerformed(ActionEvent e)
    {
        String cmd = e.getActionCommand();
        if(cmd.equals("Save"))
    {
        save();
        L.setText("System Message :O n e  R e c o r d  I n s e r t e d ");
        clear();
        b1.setEnabled(false);
    }
    if(cmd.equals("Update"))
    {
            updRec(ctr);
        L.setText("System Message :O n e  R e c o r d  U p d a t e d ");
    }
    if(cmd.equals("Cancel"))
        clear();
    if(cmd.equals("First"))
    {
        ctr=0;
        disp(ctr);
    }
    if(cmd.equals("Last"))
    {
        ctr=nor-1;
        disp(ctr);
    }
    if(cmd.equals("Prev"))
    {
        ctr--;
        if(ctr<0)
        {
            ctr = 0;
        L.setText("System Message :Sorry this is the First Record");

        }
        disp(ctr);
    }
    if(cmd.equals("Next"))
    {
        ctr++;
        if(ctr>=nor)
        {
            ctr = nor-1;
            L.setText("System Message :Sorry this is the Last Record");
        }
```

```
            disp(ctr);
        }
    }
    //<applet code=Focus height=400 width=600></applet>
```

COMPILE:

```
C:\>javac Focus.java
```

EXECUTE:

```
C:\>appletviewer Focus.java
```

OUTPUT:

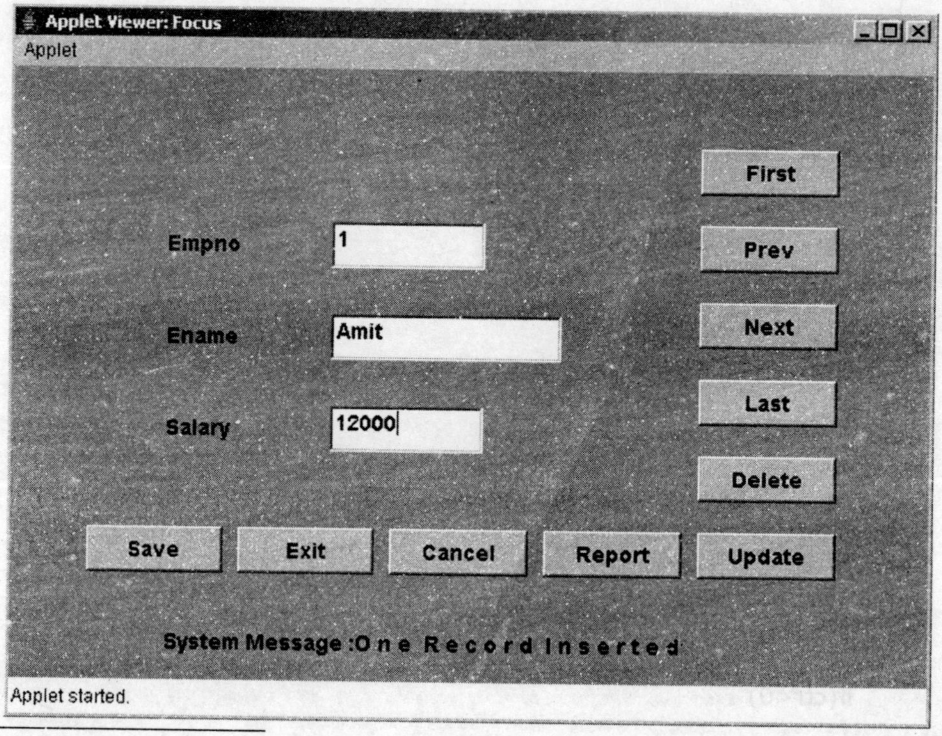

SUMMARY

— It this chapter we tried to explain all the event handling concept through examples.

— Action listener and after listeneris basically interface which are usedtop handle the events.

— In Java JDK 1, 2 also added some adapter classes, whichare subtitute ofthose interfaces which contain more methods to implement in derived classes.

EXERCISES

Q.9.1. What is Event Source ?

Q.9.2. What do you mean by Event Lintener ?

Q.9.3. Write down themethods of Mons Listener.

Q.9.4. What is the benefits of adapter classes ?

Q.9.5. Write down the methods of window Listener ?

Q.9.6. What is the difference between Mouse Listenerand Mouse Motion Listener ?

Q.9.7. Write a program to add a Button Control "Message" to display a welcome message on status bar.

Q.9.8. Write a program to add some radio button controls, which input for category such as (Gen, OBC, SC/ST) and display the selected category on the lebel.

Q.9.9. Write program to creater calculatior with proper event handling.

Q.9.10. Write a program to create a game of tick-tak-teo with Button control.

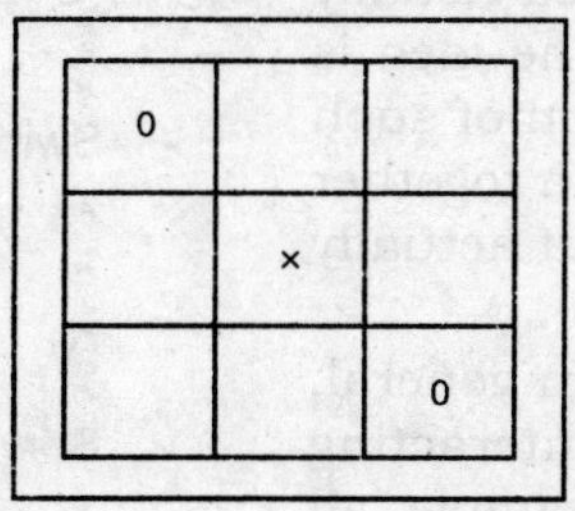

□□□

Chapter **10**

Advanced Concepts of JAVA

10.1. BEANS

A Java Bean is a reusable software component that can be manipulated visually in an application builder tool. The idea is that one can start with a collection of such components, and quickly tide them together to form complex programs without actually writing any new code.

Software components must, in general, adopt standard techniques for interacting with the rest of the world. For example, all GUI components inherit the **java.awt.Component** class, which means that one can rely on them to have certain standard methods like *paint()*, *setSize()*, etc. Java Beans are not actually required to inherit a particular base class or implement a particular interface. However, they do provide support for some or all of the following key features

- Support for *introspection*. Introspection is the process by which an application builder discovers the properties, methods and events that are associated with a Java Bean.
- Support for *properties*. These are basically member variables that control the appearance or behavior of the Java Bean.
- Support for *customization* of the appearance and behavior of a Java Bean.
- Support for *events*. This is a mechanism by which Java Beans can communicate with one another.

CHAPTER OUTLINE

- Beans
  - The BeanBox
  - Creating Beans
  - Support for Properties and Events
- Swings
  - Main Features
  - Swing Model/view design
  - Java Swing class hierarchy
  - Java Swing Examples
- Servlets
  - Why Use Servlets?
  - Perl & ASP
  - How Do We Use Servlets?

- Support for *persistent storage.* Persistence refers to the abililty to save the current state of an object, so that it can be restored at a later time.

10.1.1. The BeanBox

This is a basic tool that Sun provides for testing Java Beans. To run the BeanBox, your computer needs to have access to a **BDK** installation. To run the BeanBox, go to the *beans/beanbox* subdirectory and then type *run* on console. This will bring up three windows:

- The ToolBox window gives you a palette of sample Java Beans to choose from.
- The BeanBox window is a container within which you can visually wire beans together.
- The Properties window allows you to edit the properties of the currently selected Java Bean.

Try a simple example: choose *Juggler* bean from the ToolBox and drop an instance in the BeanBox window. Also create two instances of *OurButton.* Edit the labels of the buttons to read *start* and *stop* using the Properties window. Now wire the start button to the juggler as follows. Select the start button, then go to Edit | Events | action | actionPerformed. Connect the rubber band to the juggler. You will now see an EventTargetDialog box with a list of *Juggler* methods that could be invoked when the start button is pressed (these are the methods that either take an ActionEvent as an argument or have no arguments at all.) Choose *startJuggling* as the target method and press OK. The BeanBox now generates an adaptor class to wire the start button to the juggler. Wire the stop button to the juggler's *stopJuggling* method in a similar manner.

Now that the program has been designed, you can run it within the BeanBox. Simply press the start button to start juggling and press the stop button to stop juggling. If you wish, you can turn your program into an applet by choosing File | MakeApplet in the BeanBox. This will automatically generate a complete set of files for the applet, which can be run in the appletviewer. (Do not expect current versions of Netscape and Internet Explorer to work with this applet.)

Let's take a closer look at how the BeanBox works. On start up, it scans the directory *beans/jars* for files with the .jar extension that contain Java Beans. These beans are displayed in the ToolBox window, from where they can be selected and dropped into the BeanBox window. Next, we edited the labels of the two instances of *OurButton.* The BeanBox determined that *OurButton* has a member named *label* by looking for setter and getter methods that follow standard naming conventions called *design patterns.* If you look at the source code is given below, you will see that OurButton has two methods named

```
public void setLabel(String newLabel) {
    ...
}
public String getLabel() {
    ...
}
```

Design patterns are an implicit technique by which builder tools can introspect a Java Bean. There is also an explicit technique for exposing properties, methods and events. This involves writing a bean information class, which implements the *BeanInfo* interface.

When we wired the start button to the juggler, the BeanBox set up the juggler to respond to action events generated by the start button. The BeanBox again used design patterns to determine the type of events that can be generated by an *OurButton* object. The following design patterns indicate that *OurButton* is capable of firing *ActionEvents*.

```
public synchronized void addActionListener(ActionListener l) {
    ...
}
public synchronized void removeActionListener(ActionListener l) {    ...
}
```

By choosing Edit | Events | action | actionPerformed to connect the start button to the juggler, we were really registering an *ActionListener* with the start button. The *Juggler* bean itself is does not implement the *ActionListener* interface. Instead the BeanBox generated an event hookup adaptor, which implements *ActionListener* and simply calls the juggler's *startJuggling* method in its *actionPerformed* method:

```
// Automatically generated event hookup file.

package tmp.sunw.beanbox;
import sunw.demo.juggler.Juggler;
import java.awt.event.ActionListener;
import java.awt.event.ActionEvent;

public class ___Hookup_1474c0159e implements
    java.awt.event.ActionListener, java.io.Serializable {

    public void setTarget(sunw.demo.juggler.Juggler t) {
        target = t;
    }
    public void actionPerformed(java.awt.event.ActionEvent arg0) {

        target.startJuggling(arg0);
    }
        private sunw.demo.juggler.Juggler target;
    }
```

A similar event hookup adaptor was generated when we wired the stop button to the juggler's *stopJuggling* method.

Why not make *Juggler* implement the *ActionListener* interface directly? This is mainly a matter of convenience. Suppose that *Juggler* implemented *ActionListener* and that it was registered to receive *ActionEvents* from both the start button and the stop button. Then the *Juggler*'s *actionPerformed* method would need to examine incoming events to determine the event source, before it could know whether to call *startJuggling* or *stopJuggling*.

10.1.2. Creating a Java Bean

This example illustrates how to create a simple Java Bean. Java Bean classes must be made serializable so that they support persistent storage. To make use of the default serialization capabilities in Java, the class needs to implement the *Serializable* interface or inherit a class that implements the *Serializable* interface. Note that the *Serializable* interface does not have any methods. It just serves as a flag to say that the designer has tested the class to make sure it works with default serialization. Here is the code:

SimpleBean.java

```
import java.awt.*;
import java.io.Serializable;
public class SimpleBean extends Canvas implements Serializable {
   //Constructor sets inherited properties
   public SimpleBean() {
      setSize(60,40);
      setBackground(Color.red);
   }
}
```

Since this class extends a GUI component, *java.awt.Canvas*, it will be a visible Java Bean. Java Beans may also be invisible.

Now the Java Bean must be compiled and packaged into a JAR file. First run the compiler:

```
javac SimpleBean.java
```

Then create a manifest file

manifest.tmp

```
Name: SimpleBean.class
Java-Bean: True
```

Finally create the JAR file:

```
jar cfmv SimpleBean.jar manifest.tmp SimpleBean.class
```

The JAR file can now be placed in the beans/jars *so that the BeanBox will find it on startup, or it can be loaded subsequently by choosing File | LoadJar.*

10.1.3. Support for Properties and Events

This example builds on the SimpleBean class. It illustrates how to add customizable properties to a Java Bean and how to generate and receive property change events.

SimpleBean.java

```
import java.awt.*;
import java.io.Serializable;
import java.beans.*;
public class SimpleBean extends Canvas implements Serializable,
   java.beans.PropertyChangeListener {
   // Constructor sets inherited properties
   public SimpleBean() {
      setSize(60,40);
      setBackground(Color.red);
   }
   // This section illustrates how to add customizable properties to the Java Bean. The
names
```

```
// of the property setter and getter methods must follow specific design patterns that
   allow
// the BeanBox (or builder tool) to determine the name of the property variable
   upon
// introspection.
private Color beanColor = Color.green;
public void setBeanColor(Color newColor) {
   Color oldColor = beanColor;
   beanColor = newColor;
   repaint();
   // This relates to bound property support (see below).
   changes.firePropertyChange("beanColor", oldColor, newColor);
}
public Color getBeanColor() {
   return beanColor;
}
public void paint(Graphics g) {
   g.setColor(beanColor);
   g.fillRect(20,5,20,30);
}
// This section illustrates how to implement bound property support.  Bound property
// support allows other objects to respond when a property change occurs in this
   Java
// Bean.  Remember that each property setter method must fire a property change
   event,
// so that registered listeners can be properly notified.  The addPropertyChangeListener
// and removePropertyChangeListener methods follow design patterns that indicate
   the
// ability of this Java Bean to generate property change events. As it happens, these
// methods overide methods of the same names, which are inherited through
   java.awt.Canvas.
private PropertyChangeSupport changes = new PropertyChangeSupport(this);
public void addPropertyChangeListener(PropertyChangeListener l) {
   changes.addPropertyChangeListener(l);
}
public void removePropertyChangeListener(PropertyChangeListener l) {
   changes.removePropertyChangeListener(l);
}
// This section illustrates how to implement a bound property listener, which will
   allow this
// Java Bean to register itself to receive property change events fired by other
   objects.
// Registration simply involves making a call to the other object's
   addPropertyChangeListener
// method with this Java Bean as the argument.  If you are using the BeanBox,
   however, you
// will typically use the event hookup adaptor mechanism to receive the events.  In
   this case,
```

```
// you can set the target method to be the propertyChange method.
   (Another word about the
// BeanBox: the Edit | bind property option is a useful way to make a property
   change in one
// object automatically trigger a property change in another object. In this case, the
   BeanBox
// will invoke the correct property setter using code in sunw/ beanbox/
   PropertyHookup.java.
// An adaptor class will not be generated in this case.)
public void propertyChange(PropertyChangeEvent evt) {
  String propertyName = evt.getPropertyName();
  System.out.println("Received property change event " + propertyName);
 }
}
```

10.2. SWINGS

To create a Java program with a graphical user interface (GUI), you have to use Swing and AWT.

The Swing toolkit includes a rich set of components for building GUIs and adding interactivity to Java applications. Swing includes all the components you would expect from a modern toolkit: table controls, list controls, tree controls, buttons, and labels.

Swing is far from a simple component toolkit, however. It includes rich undo support, a highly customizable text package, integrated internationalization and accessibility support. To truly leverage the cross-platform capabilities of the Java platform, Swing supports numerous look and feels, including the ability to create your own look and feel. The ability to create a custom look and feel is made easier with Synth, a look and feel specifically designed to be customized.

Swing wouldn't be a component toolkit without the basic user interface primitives such as drag and drop, event handling, customizable painting, and window management.

Swing is part of the Java Foundation Classes (JFC). The JFC also include other features important to a GUI program, such as the ability to add rich graphics functionality and the ability to create a program that can work in different languages and by users with different input devices.

The following list shows some of the features that Swing and the Java Foundation Classes provide.

10.2.1. Main Features

- **Lightweight**. Not built on native window-system windows.
- **Much bigger set of built-in controls**. Trees, image buttons, tabbed panes, sliders, toolbars, color choosers, tables, text areas to display HTML or RTF, etc.
- **Much more customizable**. Can change border, text alignment, or add image to almost any control. Can customize how minor features are drawn. Can separate internal representation from visual appearance.
- **"Pluggable" look and feel**. Can change look and feel at runtime, or design own look and feel.
- **Many miscellaneous new features**. Double-buffering built in, tool tips, dockable tool bars, keyboard accelerators, custom cursors, etc.

Swing, which is an extension library to the AWT, includes new and improved components that enhance the look and functionality of GUIs. Swing can be used to build Standalone swing gui Apps as well as Servlets and Applets. It employs a model/view design architecture. Swing is more portable and more flexible than AWT.

10.2.2. Swing Model/view Design

The "view part" of the MV design is implemented with a component object and the UI object. The "model part" of the MV design is implemented by a model object and a change listener object.

Swing is built on top of AWT and is entirely written in Java, using AWT's lightweight component support. In particular, unlike AWT, t he architecture of Swing components makes it easy to customize both their appearance and behavior. Components from AWT and Swing can be mixed, allowing you to add Swing support to existing AWT-based programs. For example, swing components such as JSlider, JButton and JCheckbox could be used in the same program with standard AWT labels, textfields and scrollbars. You could subclass the existing Swing UI, model, or change listener classes without having to reinvent the entire implementation. Swing also has the ability to replace these objects on-the-fly.

MVC Architecture

- Model represents the data
- View as a visual representation of the data
- Controller takes input and translates it to changes in data

In Swing, classes that represent GUI components have names beginning with the letter J. Some examples are JButton, JLabel, and JSlider. Altogether there are more than 250 new classes and 75 interfaces in Swing — twice as many as in AWT.

10.2.3. Java Swing Class Hierarchy

The class **JComponent**, descended directly from Container, is the root class for most of Swing's user interface components.

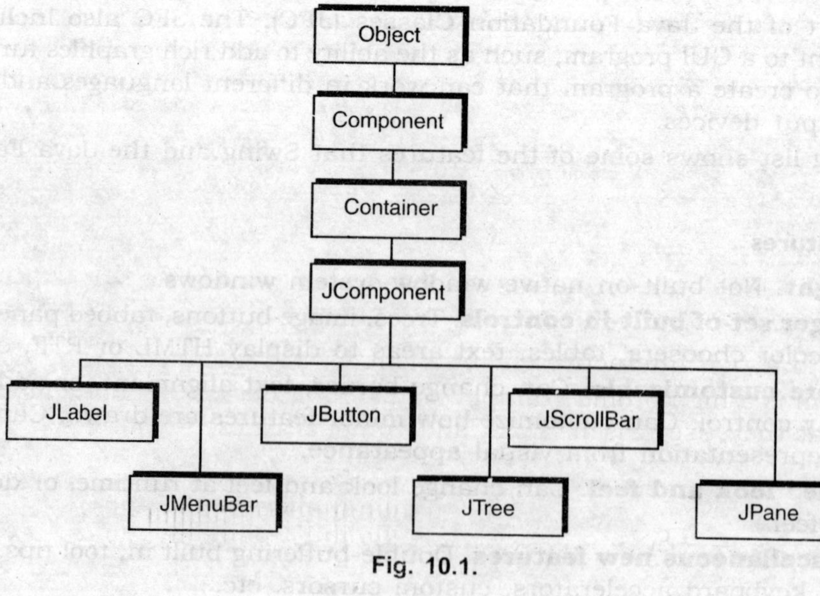

Fig. 10.1.

Swing contains **components** that you'll use to build a GUI. I am listing you some of the commonly used Swing components. To learn and understand these swing programs, AWT Programming knowledge **is not** required.

10.2.4. Java Swing Examples

Below is a java swing code for the traditional Hello World program.

Basically, the idea behind this Hello World program is to learn how to create a java program, compile and run it. To create your java source code you can use any editor(Text pad/Edit plus are my favorites) or you can use an IDE like **Eclipse**.

```
import javax.swing.JFrame;
import javax.swing.JLabel;

//import statements
//Check if window closes automatically. Otherwise add suitable code
public class HelloWorldFrame extends JFrame {
    public static void main(String args[]) {
        new HelloWorldFrame();
    }
    HelloWorldFrame(){
        JLabel jlbHelloWorld = new JLabel("Hello World");
        add(jlbHelloWorld);
        this.setSize(100, 100);
        // pack();
        setVisible(true);
    }
}
```

Output

10.3. SERVLETS

Servlets provide the means through which Web Applications can be written in Java. What are Web Applications? Web Applications allow a web site to be dynamic rather than straight static pages. In other words, Servlets transform a web site with plain text and images into a rich interactive environment for the user.

Amazon.com is a high profile example of this. Amazon has a proprietary "Web Store" application that they use to sell books and compact discs online.

The attractiveness of such applications is that they are built on the foundations of the World Wide Web. Thus, these applications can be run anywhere in the world at any time and are completely cross platform. All you need to run them is a Web Browser.

Before the Servlet API was introduced, web designers could add interactive functionality to their sites.

In the beginning, Web Servers were made to just serve basic text content and images. However, in order to provide an interactive environment, the web server had to be extended to allow programs or scripts to be executed. This interface was standardized as CGI.

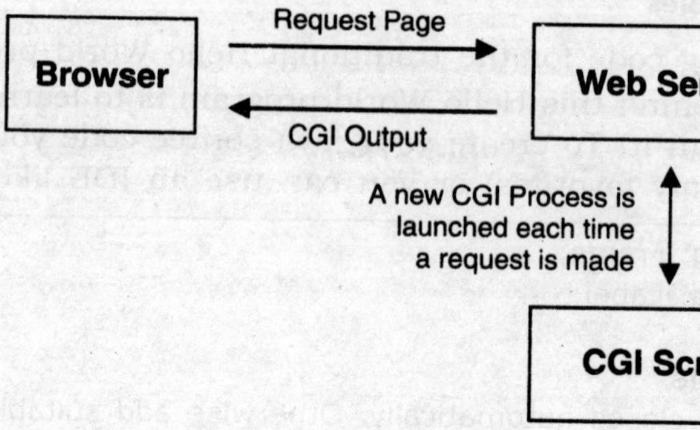

Fig. 10.2.

The browser would load an HTML page consisting of some form elements such as text boxes or check boxes. Then, the user would fill in the form and submit it to the web server. The Web Server would take the form data and submit the data to the CGI program. When the CGI program runs, it outputs the relevant HTML or other data back to the Web Server which acts as a middle-man and sends the data back to the browser. The diagram below illustrates the life cycle of a browser getting results from a CGI program.

10.3.1. Why Use Servlets?

Servlets have a variety of enhancements over plain CGI. The most fundamental advantage of Servlets is that they are tightly coupled to the Web Server. If you recall from the diagram above, CGI scripts are typically launched as a separate process by the web server whose lifetime exists only during the particular user request.

It turns out that the CGI model is extremely inefficient and slow especially with interpreted languages such as Perl or Java.

First, operating systems typically incur some overhead in simply launching another process. When you think about it, an operating system really has to do quite a bit. It has to: allocate memory for the process copy information about the environment that the parent process was running in so that there is a security context and the web server (parent process) must be constantly aware of the child process in order to send the CGI output back to the user's web browser.

Second, an interpreted language typically has to load an entire program to interpret the script before the program can even execute. In the case of Perl, this is the Perl executable. In the case of Java, it is the Java binary. Even on a small program and a relatively fast machine, a Java program can take a second or more to start.

Enter Java Servlets to the save the day. The Servlet engine runs persistently alongside the Web Server itself. First, because the Servlet engine is running constantly, the only time a new process is ever necessary if at all is when the web server starts. Second, the class files themselves can be cached inside the JVM (Java Virtual Machine), so subsequent calls to Servlets do not have to go through the process of loading them into memory and past result sets can be cached in the web server itself.

The mere placement of the Java Virtual Machine inside a Web Server was really just the first step for Java Servlets. Subsequent enhancements have been made to the API that add many value-added features. These include:

Session persistence JSPs (Java Server Pages) and external utility libraries to take advantage of the caching nature of Servlets for resources such as handles to relational databases.

The following is a brief listing of the major features of the current **http://www.javasoft.com/products/servlet/index.html:**

- Persistent Java Engine
- Cached Classes
- Cached Resources (eg Database handles)
- Session handling (w/cookies and/or URL rewriting)
- Java Server Pages
- Servlet Chaining
- Inter-Servlet Communication
- HTTP Header Integration
- Internationalization features for world-wide use
- Support for Distributed Servlets
- Support for Web Application Archives (WAR files)
- Response Buffering

10.3.2. Perl & ASP

Although this book if about Java, it is difficult to escape the notion that other established web technologies can help you write interactive web sites. Perl is the foundation upon which many dynamic pages have been written, and Active Server Pages (ASP) from Microsoft is a close contender for Microsoft IIS Servers.

It turns out that these other technologies do have a place in the world of Web Programming. And indeed, they can provide many of the same features of Java Servlets.

http://www.perl.com/ has been the stable of many webmasters since CGI was first implemented in Web Servers. Perl as a language is extremely easy to use and learn because it is extremely forgiving of syntax issues. The main downside for Perl is that CGI scripts written in Perl tend to be slow. There are accelerators available for both **http://www.activestate.com/** and UNIX Web servers such as **http://perl.apache.org/** . However, most of these accelerators are not yet cross platform.

http://msdn.microsoft.com/workshop/server/asp/ASPover.asp are very similar to Perl in that the language (VBScript) is extremely easy to learn. ASPs are also written as snippets of code inside of HTML which makes it intuitive for web designers and HTML coders to pick up. In addition, they support sessions just like Java Servlets. However, the downside to ASPs is that they are generally limited to Microsoft IIS and PWS Web Servers.

The cool thing about Servlets is that you will get the advantages of a persistent Java engine on just about whatever platform you wish to run on, whether it is Windows or a flavor of UNIX. In addition, you will get additional features such as session management built in for you.

10.3.3. How Do We Use Servlets?

There are usually two steps to installing Servlet support on your Web Server. First, you need to get a Servlet engine. There are a variety of them out there, and here are but a few:

- **http://www1.allaire.com/products/jrun/**
- **http://www.newatlanta.com/**
- and JavaSoft has their own **http://www.sun.com/software/jwebserver/index.html**

Second, if the current version of the Java Development Kit (JDK) or Java Runtime Engine (JRE) is not bundled with the servlet engine you download, you need to download the latest copy from the **http://www.javasoft.com/** web site.

SUMMARY

- Java Beans is a reusable software components that can be manipulated in builder tools.
- Java Beans obtains all the benefits of Java's "write-once, run-anywhere" paradigm.
- Swing is a GUI tool-kit for Java
- In general you only have a add a "**J**" in front of the class name of simple AWT components do part them into SWING components.
- Servlets are modulus that run inside request/response-oriented servers, such as Java-enabled web servers.
- Servlets can handle multiple requests concurrently, they can synrhronize requests to support systems such as online conferencing.

SOLVED PROBLEMS

Q.10.1. What is JavaBeans?

Ans. JavaBeans is a portable, platform-independent component model written in the Java programming language, developed in collaboration with industry leaders. It enables developers to write reusable components once and run them anywhere -- benefiting from the platform-independent power of Java technology. JavaBeans acts as a Bridge between proprietary component models and provides a seamless and powerful means for developers to build components that run in ActiveX container applications.

Q.10.2. What is a Bean? Why isn't a Bean an Applet?

Ans. JavaBeans components, or Beans, are reusable software components that can be manipulated visually in a builder tool. Beans can be combined to create traditional applications, or their smaller web-oriented brethren, applets. In addition, applets can be designed to work as reusable Beans.

Individual Beans will function quite differently, but typical unifying features that distinguish a Bean are:

- **Introspection:** enables a builder tool to analyze how a Bean works
- **Customization:** enables a developer to use an app builder tool to customize the appearance and behavior of a Bean
- **Events:** enables Beans to communicate and connect together
- **Properties:** enable developers to customize and program with Beans

- **Persistence:** enables developers to customize Beans in an app builder, and then retrieve those Beans, with customized features intact, for future use

Q.10.3. Why are component architectures useful?

Ans. Developers are turning to creating components rather than monolithic applications to free themselves from slow, expensive application development, and to build up a portable, reusable code base. This enables developers to quickly attack new market opportunities, new joint development opportunities, and new ways to sell smaller packages of software.

Q.10.4. Is JavaBeans a complete component architecture?

Ans. JavaBeans is a complete component model. It supports the standard component architecture features of properties, events, methods, and persistence. In addition, JavaBeans provides support for introspection (to allow automatic analysis of a JavaBeans component) and customization (to make it easy to configure a JavaBeans component).

Q.10.5. Why a component architecture for the Java platform?

Ans. JavaBeans brings the extraordinary power of the Java platform to component development, offering the ideal environment for a developer who wants to extend the concept of reusable component development beyond one platform and one architecture to embrace every platform and every architecture in the industry.

Q.10.6. What is the relationship between Sun's JFCs and JavaBeans?

Ans. The JFC (Java Foundation Classes) is based upon the AWT (Abstract Windowing Toolkit), which has been part of the Java platform from the beginning. JFC effectively adds a richer set of visual elements for building JavaBeans components and applications. See the JFC web site for more information.

Q.10.7. What are the security implications for downloading Beans over the Internet?

Ans. JavaBeans does not add any security features to the Java platform. Rather, JavaBeans components have full access to the broad range of security features that are part of the Java platform. JavaBeans components can be used to build a range of different kinds of solutions from full-fledged Java desktop applications to web-based Applets.

Q.10.8. Are JavaBeans a standard?

Ans. JavaBeans are a public API published by Sun. However, JavaBeans have been neither approved by, nor are under the control of, any independent standards body.

Q.10.9. On which platforms can we use JavaBeans?

Ans. Because JavaBeans are just Java classes written to a specific standard, you can use them on any platform for which there is a Java Virtual Machine that supports JDK 1.1-compatible programs. This means that Windows, Macintosh, and most flavors of Unix can run Java programs that take advantage of JavaBeans.

Q.10.10. Do we need special software to create or use JavaBeans?

Ans. Although you don't need special tools to create JavaBeans, software that is JavaBean-aware will probably make the process easier. In addition, some development environments, especially those that allow for WYSIWYG layout of controls, need to be specifically aware of JavaBeans in order for you to be able to use JavaBean components similarly to built-in controls. In fact, using JavaBeans in an IDE that supports them is as easy as dragging a JavaBean from a component palette into your project in a way similar to that popularized by Microsoft Visual Basic. For example, both Symantec's Visual Café and Borland/Inprise's JBuilder make working with and creating JavaBeans a breeze. Other tools, such as Microsoft's Visual J++, IBM's VisualAge for Java, and Sybase's PowerJ also suppor. JavaBeans to varying degrees.

Q.10.11. What's the difference between Java applets and JavaBean components?

Ans. Both JavaBeans and applets are Java classes, but they are designed with different purposes in mind. An applet is a mini-application that is designed to run inside a browser, whereas a JavaBean is a component coded to a specific standard that is designed to be used within an application or an applet. While it is possible to create a wrapper to turn an applet into a JavaBean, or vice-versa, this conversion is usually unnecessary because the two serve different purposes.

Q.10.12. If I write a program that uses JavaBeans how do we make sure that our users have the right components on their machine?

Ans. You can package one or more JavaBeans into a special JAR file that the Java class loader will automatically download to the user's machine if needed. JAR files are ZIP files that you can use to compress one or more related JavaBeans and any associated files for downloading to a user's machine as a package. JAR files differ from regular ZIP files only in that they must include a file named MANIFEST that acts as the JAR's table of contents.

Q.10.13. What are the differences between JavaBeans and Enterprise JavaBeans?

Ans. JavaBeans were originally designed for user-interface and client-side tasks. To allow for server-side, component-based development in Java, Sun has released the Enterprise JavaBeans (EJB) specification that adds server-side features such as transactions, persistence, and scalability.

Q.10.14. What is Swing?

Ans. According to Sun, the correct name is "JFC Project Swing". Everyone else refers to it as simply "Swing". Swing is a set of lightweight (implemented completely in Java) graphical components. It includes not only the equivalents of AWT components, but also new ones such as table and tree widgets.

Swing is a standard part of Java 2 and higher. Swing 1.1.1 is the last version of Swing which will be compatible with JDK 1.1.

Q.10.15. What is the JFC?

Ans. JFC stands for "Java Foundation Classes". This is the term for several capabilities included in the Java 2 Platform. (Some pieces are available for JDK 1.1.)

- Swing components
- Accessibility
- Pluggable look-and-feel
- Java 2D (JDK 1.2 only)
- Drag-and-drop (JDK 1.2 only)

Q.10.16. Can you run Swing under a browser?

Ans. Swing is not included by default in current versions of Netscape Navigator or Microsoft Internet Explorer. See the README or CHANGES.TXT files that come in the Swing distribution for an explanation of how to make Swing work with the browser. Alternatively, the Java Plugin includes Swing, and works with even the older 3.x versions of these browsers. JDK 1.2.2 is now included on all of those AOL coasters you get, too.

OS/2's version of Netscape uses the system's default Java implementation, so Swing works there. All versions of Mozilla (including Netscape 6.0) will behave this way in the future.

Q.10.17. What package is Swing in?

Ans. Swing has finally settled into the `javax.swing` package. Swing 1.0.3 was the last version to use the old package names.

Q.10.18. What is Model-View-Controller? What do design patterns have to do with Swing?

Ans. Design patterns are a literary form that document tried-and-true solutions to design problems. A good starting point is the classic book *Design Patterns* by Gamma et al. The most famous design pattern is "Model-View-Controller" (also known as "Observer"); this pattern was popularized in Smalltalk. MVC addresses the problem of how to build software for user interfaces. It says to divide the software into three parts: the Model is the basic behavior of the system, independent of any user interface. The View is a presentation of the Model, and the Controller determines how the user interacts with the Model. The View and Controller know about the Model, but the Model is almost unaware of their existence: it merely knows that if it changes, it has to send a message to any registered listener warning it that there have been changes.

The easiest place to see this approach is in `JTable`. You can define a `TableModel` object (the Model), and tell a `JTable` (combined View-Controller) to observe that model. When the data changes, the `JTable` is notified, and it updates its display. Several Views could monitor a single table; the Model notifies them of changes, and lets the Views figure out how to update the display. (A common mistake is to write code to update the `TableModel` but not send any notifications. This shows up as the changes not being reflected unless you do a resize or force a repaint. Notification is part of the Model's job. `DefaultTableModel` sends these notifications for you, so see its source code for examples of what to do)

There are other design patterns used in the Swing libraries. For example, the "Composite" pattern describes how to handle a containment hierarchy of objects; `JComponent` implements this pattern. "Decorator" describes how to compose functionality; using CompoundBorder to nest one border inside another uses this approach. Time spent studying design patterns will often help you better understand the design of Swing.

Q.10.19. Explain the life cycle methods of a Servlet.

Ans. The javax.servlet.Servlet interface defines the three methods known as life-cycle method.public void init(ServletConfig config) throws ServletExceptionpublic void service(ServletRequest req, ServletResponse res) throws ServletException, IOExceptionpublic void destroy()First the servlet is constructed, then initialized wih the init() method.Any request from client are handled initially by the service() method before delegating to the doXxx() methods in the case of HttpServlet.The servlet is removed from service, destroyed with the destroy() methid, then garbaged collected and finalized.

Q.10.20. What is the difference between the getRequestDispatcher(String path) method of javax.servlet.ServletRequest interface and javax.servlet.ServletContext interface?

Ans. The getRequestDispatcher(String path) method of javax.servlet.ServletRequest interface accepts parameter the path to the resource to be included or forwarded to, which can be relative to the request of the calling servlet.If the path begins with a "/" it is interpreted as relative to the current context root.The getRequestDispatcher(String path) method of javax.servlet.ServletContext interface cannot accepts relative paths.All path must sart with a "/ " and are interpreted as relative to curent context root.

Q.10.21. Explain the directory structure of a web application.

Ans. The directory structure of a web application consists of two parts.A private directory called WEB-INFA public resource directory which contains public resource folder.WEB-INF folder consists of 1.web.xml2.classes directory3.lib directory

Q.10.22. What are the common mechanisms used for session tracking?

Ans. CookiesSSL sessionsURL-rewriting

Q.10.23. Explain ServletContext.

Ans. ServletContext interface is a window for a servlet to view it's environment.A servlet can use this interface to get information such as initialization parameters for the web applicationor servlet container's version.Every web application has one and only one ServletContext and is accessible to all active resource of that application.

Q.10.24. What is the difference between Difference between doGet() and doPost()?

Ans. A doGet() method is limited with 2k of data to be sent, and doPost() method doesn't have this limitation. A request string for doGet() looks like the following: http://www.allapplabs.com/svt1?p1=v1&p2=v2&...&pN=vNdoPost() method call doesn't need a long text tail after a servlet name in a request.All parameters are stored in a request itself, not in a request string, and it's impossible to guess the data transmitted to a servlet only looking at a request string.

Q.10.25. What is the difference between HttpServlet and GenericServlet?

Ans. A GenericServlet has a service() method aimed to handle requests.HttpServlet extends GenericServlet and adds support for doGet(), doPost(), doHead() methods (HTTP 1.0) plus doPut(), doOptions(), doDelete(), doTrace() methods (HTTP 1.1).Both these classes are abstract.

Q.10.26. What is Servlet?

Ans. A servlet is a Java technology-based Web component, managed by a container called servlet container or servlet engine, that generates dynamic content and interacts with web clients via a request\/response paradigm.

Q.10.27. Why is Servlet so popular?

Ans. Because servlets are platform-independent Java classes that are compiled to platform-neutral byte code that can be loaded dynamically into and run by a Java technology-enabled Web server.

Q.10.28. What is servlet container?

Ans. The servlet container is a part of a Web server or application server that provides the network services over which requests and responses are sent, decodes MIME-based requests, and formats MIME-based responses. A servlet container also contains and manages servlets through their lifecycle.

Q.10.29. When a client request is sent to the servlet container, how does the container choose which servlet to invoke?

Ans. The servlet container determines which servlet to invoke based on the configuration of its servlets, and calls it with objects representing the request and response.

Q.10.30. If a servlet is not properly initialized, what exception may be thrown?

Ans. During initialization or service of a request, the servlet instance can throw an UnavailableException or a ServletException.

Q.10.31. Given the request path below, which are context path, servlet path and path info?

Ans./bookstore/education/index.html

context path:/bookstore

servlet path:/education

path info: /index.html

Q.10.32. What is filter? Can filter be used as request or response?

Ans. A filter is a reusable piece of code that can transform the content of HTTP requests,responses, and header information. Filters do not generally create a response or respond to a request as servlets do, rather they modify or adapt the requests for a resource, and modify or adapt responses from a resource.

Q.10.33. When using servlets to build the HTML, you build a DOCTYPE line, why do you do that?

Ans. I know all major browsers ignore it even though the HTML 3.2 and 4.0 specifications require it. But building a DOCTYPE line tells HTML validators which version of HTML you are using so they know which specification to check your document against. These validators are valuable debugging services, helping you catch HTML syntax errors.

Q.10.34. What is new in ServletRequest interface ? (Servlet 2.4)

Ans. The following methods have been added to ServletRequest 2.4 version:

```
public int getRemotePort()
public java.lang.String getLocalName()
public java.lang.String getLocalAddr()
public int getLocalPort()
```

Q.10.35. Request parameter How to find whether a parameter exists in the request object?

Ans. 1. boolean hasFoo = !(request.getParameter("foo") == null || request.getParameter("foo").equals(""));

2. boolean hasParameter = request.getParameterMap().contains(theParameter); (which works in Servlet 2.3+)

Q.10.36. How can I send user authentication information while making URL Connection?

Ans. You'll want to use HttpURLConnection.setRequestProperty and set all the appropriate headers to HTTP authorization.

Q.10.37. Can we use the constructor, instead of init(), to initialize servlet?

Ans. Yes, of course you can use the constructor instead of init(). There's nothing to stop you. But you shouldn't. The original reason for init() was that ancient versions of Java couldn't dynamically invoke constructors with arguments, so there was no way to give the constructur a ServletConfig. That no longer applies, but servlet containers still will only call your no-arg constructor. So you won't have access to a ServletConfig or ServletContext.

Q.10.38. How can a servlet refresh automatically if some new data has entered the database?

Ans. You can use a client-side Refresh or Server Push

Q.10.39. The code in a finally clause will never fail to execute, right?

Ans. Using System.exit(1); in try block will not allow finally code to execute

Q.10.40. What mechanisms are used by a Servlet Container to maintain session information?

Ans. Cookies, URL rewriting, and HTTPS protocol information are used to maintain session information

Q.10.41. Difference between GET and POST

Ans. In GET your entire form submission can be encapsulated in one URL, like a hyperlink. query length is limited to 260 characters, not secure, faster, quick and easy.

In POST Your name/value pairs inside the body of the HTTP request, which makes for a cleaner URL and imposes no size limitations on the form's output. It is used to send a chunk of data to the server to be processed, more versatile, most secure.

Q.10.42. What is session?

Ans. The session is an object used by a servlet to track a user's interaction with a Web application across multiple HTTP requests.

Q.10.43. What is servlet mapping?

Ans. The servlet mapping defines an association between a URL pattern and a servlet. The mapping is used to map requests to servlets.

Q.10.44. What is servlet context ?

Ans. The servlet context is an object that contains a servlet's view of the Web application within which the servlet is running. Using the context, a servlet can log events, obtain URL references to resources, and set and store attributes that other servlets in the context can use. (answer supplied by Sun's tutorial).

Q.10.45. Which interface must be implemented by all servlets?

Ans. Servlet interface.

Q.10.46. Explain the life cycle of Servlet.

Ans. Loaded(by the container for first request or on start up if config file suggests load-on-startup), initialized(using init()), service(service() or doGet() or doPost()..), destroy(destroy()) and unloaded.

Q.10.47. When is the servlet instance created in the life cycle of servlet? What is the importance of configuring a servlet?

Ans. An instance of servlet is created when the servlet is loaded for the first time in the container. Init() method is used to configure this servlet instance. This method is called only once in the life time of a servlet, hence it makes sense to write all those configuration details about a servlet which are required for the whole life of a servlet in this method.

Q.10.48. Why don't we write a constructor in a servlet?

Ans. Container writes a no argument constructor for our servlet.

EXERCISES

Q.10.1. What do you mean by the JavaBeans?
Q.10.2. What are the advantages of Java Beans?
Q.10.3. What do you mean by reusable components?
Q.10.4. What is the use of JAR Files? How to create a JAR file?
Q.10.5. What is Swing ? Give the name of the Swing package.
Q.10.6. What is Model-view-controller ?
Q.10.7. What do you mean by by the JAVA Servlets ?
Q.10.8. Explain the architecture of Servlet.
Q.10.9. Explain the life cycle of Servlet.

□□□

Chapter 11

Website Designing

CHAPTER OUTLINE

- Internet
    - Internet Services
- Email
- HTML
    - Simple rules of HTML
    - HTML Tags
- Java Applets
    - Java Applet Vs Applications
    - Restrictions on a Java Applet
    - The structure of Java Applet
- The Applet Class
    - The Life and Death of an Applet
    - Example Applets
- JSP
- JSP Tags
    - Declaration Tag (<%! %>)
    - Expression Tag (<% = %>)
    - Directive Tag (<%@ directive ... %>)
    - Scriptlet Tag (<% ... %>)
    - Action Tag
- Creating a Form

11.1. INTERNET

The internet is made up of millions of computers linked together around the world in such a way that information can be sent from any computer to any other 24 hours a day. These computers can be in homes, schools, universities, government departments, or businesses small and large. They can be any type of computer and be single personal computers or workstations on a school or a company network. The internet is often described as 'a network of networks' because all the smaller networks of organisations are linked together into the one giant network called the internet. All computers are pretty much equal once connected to the internet, the only difference will be the speed of the connection which is dependent on your Internet Service Provider (ISP) and your own modem.

> **Concept**
>
> *Internet is a network of computer networks which operates world-wide using a common set of communication protocols. It is a vast collection of inter-connected networks across the world that all use the TCP/IP protocols.*

The internet has developed a very strong community base where information, software and expert advice are freely shared and for this reason users have developed a very strong protective stance on freedom of speech, freedom from commercial interests, and unsuitable material on the web.

11.1.1. Internet Services

There are so many things you can do and participate in once connected to the internet. They include using a range of services to communicate and share information and things quickly and inexpensively with tens of millions of people, both young and old and from diverse cultures around the world. For example:

- You'll be able to keep in touch and send things to colleagues and friends using electronic mail, internet telephone, keyboard chat and video conferencing.
- You can also tap into thousands of databases, libraries and newsgroups around the world to gather information on any topics of interest for work or recreation. The information can be in the form of text, pictures or even video material.
- This means you can stay up to date with news, sports, weather and any current affairs around the world with information updated daily, hourly or instantly.
- You can also locate and download computer software and other products that are available in cyberspace
- You can listen to sounds and music, and watch digital movies,
- There are also a growing number of interactive multimedia games and educational tools.

And as well as using the Internet for receiving things you will be able to publish information about your school or hobbies.

11.2. ELECTRONIC MAIL

Electronic mail, or Email, is the term used to describe the tool which allows one computer user to send a message to one or more other computer users over a computer network in a digital form. Email can be sent internally to members of an organisation through their internal computer network or they can be sent externally to anyone in the world by using the network of computer networks that make up the internet.

Concept

E-mail is a system whereby a computer user can exchange messages with other computer users (or group of users) via a communications network.

It's very similar to our traditional mail system with post boxes, post offices, envelopes and addresses. The difference is that instead of Australia Post delivering your messages around the globe for you, messages are sent electronically to other computer users via the computer network (the internet) taking a fraction of the time to get there AND, you don't have to walk to the letter box!

Just as you can send attachments with your letters in the normal post (snail mail) you can also send computer documents, graphics, software or anything else that can be turned into a digital form attached to an email message.

11.2.1. How does it Work?

In order for messages to be sent from one computer to another, your message needs to be converted into a digital form and forwarded to a computer that acts as a mail server or post office. This mail server sorts and directs your mail for you. The only way this mail server can direct mail though, to all users is by being connected to a network that all the users are also connected to.

This network can be internal (a stand-alone network) which means you can only send email to other users on that network. If your mail server is connected to the Internet you can also send your email messages to any other computer user that is connected to the internet anywhere in the world because the internet is a network of all the little networks of organisations around the world.

This mail server can be within your organisation or with an Internet Service Provider, so you would connect to it by logging into your email account. When you send your email message, the mail server decides whether the message is to be passed on to a user on its immediate network or it will pass it onto another mail server on another network closest to it through these combination of networks and each mail server keeps passing it on until it reaches its intended destination. This is known as the 'store' and 'forward' system, storing your message at various points on the path to its receiver waiting for the link to be free for it to forward your message on the next part of its journey.

The digital data can be broken up and follow different paths (go through different mail servers) to gets to its destination but always meets up when it gets to the post office or mail server of the recipient. The mail will stay at the post office until the recipient decide to collect it which is done by logging into their email account.

11.2.2. Requirements for e-mail

Email Software

In order to read or send any electronic mail you will need special software.

Microsoft Outlook is one of the most commonly used software packages and comes with windows operating system. If you have several e-mail or newsgroup accounts, you can work with all of them within one window. You can also create multiple users, or identities, for the same computer. Each identity has unique e-mail folders and an individual Address Book. Multiple identities make it easy for you to keep work e-mail separate from personal e-mail, and also to keep individual users' e-mail separated.

Web browsers like Netscape and Explorer now have an email function also but they are not as sophisticated and reliable. One benefit of this type of email package is that where addresses of World Wide Web pages are given in the body of the message you can click on the address and the browser will automatically display the page.

Email Account

To become an email user you need to have an email account set up for you by an Internet Service Provider or your organisation. This would normally occur automatically when you get your internet connection if you're setting up your account privately. With a school or business you would normally gain access to email by talking to your technical advisor where a user account will be set up for you. When you get your account you will be given an email address, a Username and a password which you would use to access your email.

You can obtain free email accounts through certain web sites on the internet (such as Yahoo, MSN, Hotmail, Rediffmail, etc.). You check your mail through your web browser by looking at a web page, typing in your name and password. These types of services are nowhere near as secure as having your own email software on your own computer and can be slow with delays. They are free though and provide additional services such as receiving regular news from various sites of your choosing.

Email Address

Each email user is given an email address which distinguishes them from each other like your name and street address distinguish your house from your neighbors. Your email address is usually your Username or User ID, plus the domain name of the computer through which you normally gain access to the internet with an @ symbol in-between.. - (seemakapoor@hclinfo.com). As you can see here Username is 'seemakapoor' and the domain name for the company is 'hclinfo', where seema is working. Your Username doesn't necessarily have to be your real name it can be a variation or any name of your choosing - eg. (wizard@hclinfo.com).

With free email accounts like Hotmail or Yahoo you will get its own domain name eg. neeraj_2000@hotmail.com, nehajain@yahoo.com.

11.2.3. Understanding Domain Names

A domain name is the Internet address of a computer. All computers connected to the internet have a domain name. Your computer itself may not have a domain name but the computer that you connect to firstly to enable you to gain access to the internet will have a domain name. For example you may connect to the internet through a computer which is called a server belonging to your organisation or through your Internet Service Providers server. By having a domain name all computers on the internet are able to locate each other and forward messages and requests.

When you specify an email address or IP address of a website it accesses the server/ computer by the domain name that you specify by accessing a database known as a DNS Name Server. You may have seen the error messages "No DNS Entry" before when typing in a web site address. This means it is unable to locate the domain address of the computer where the web page is that you specified.

A domain name is separated by periods (.) and will normally include the name of the organisation (or an acronym of it), for example the domain name for HCL is www.hclinfo.com.

The email system at the HCL uses names such as **hclinfo** followed by **@hclinfo.com** which is the domain name of the server/computer that forwards electronic mail to each user on the network.

11.2.4. Interpreting Email Addresses

Standard naming rules are used for domain names which show the type of organisations and country the people have internet accounts with, and because of this you can remember email addresses as well as web site addresses with greater ease. You can also work out where a person comes from by looking at the email address.

For example, if you received an email message from J.Smith@zoo.org.au you could tell from the domain name that this message came from Australia (AU) from a non-profit organisation (ORG) which is the Zoo which is an acronym for Melbourne Zoo.

Here is a sample of some of the standard naming rules used.

Name	Description
com	Commercial Organisations
edu	Educational Institutions

org	Non-profit organisations
net	Networks
gov	Non-military government organisations
mil	MIlitary government organisations
int	International organisations

Name	Description
au	Australia
uk	United Kingdom
fr	France
in	india

In some cases email addresses do not include a country code. This is often the case with email addresses of users from the United States but this is also becoming more common in other countries.

11.2.5. Sending Email

Email messages are often less formal than letters and can sometimes be as short as one single word. Because it is treated more casually than a formal letter sometimes people aren't as careful with what they say in an email message.

Recipients Name & Address

You can't send email without having the other person's full email address. Without it its like mailing an envelope without any address on it. Just as you would put the person's name and address on the envelope, you would put their email address, comprised of their Username and domain name, on the email message. You can include their real name in the message also.

Most email packages allow you to set up an email address book so that you only need to click on the person you would like to send the message to once you have put them in your book. All you need to do is use the automatic addressing feature to transfer the address to your new message.

As you would normally put your return name and address on the back of an envelope so the message can be returned to you if there is a problem - you can include this on your email software when you set it up so it does the same with the email messages you send.

Subject

If you type in the subject, this subject header will show up on their email program so that they get an idea of what your message is about - you can type in Urgent! to let the recipient know to open it right away.

Time & Date

The email program automatically puts in the time, the date.

Main Body

This is where you type the main body of your message.

Attachments

Documents can also be attached to email messages. To attach a document or other type of digital file to an outgoing message, select Attach File. A standard dialogue box will be displayed. Select the document you want to send and click OK. The name of the file will appear after the Attachments: field of the header.

CC (carbon copy)

You can forward a copy of your message to the user you include here. All you need to do is include their email address in this section.

BCC (blind carbon copy)

This field allows you to send a copy of a message to someone else without the original recipient knowing.

11.2.6. Checking for Email

Inbox

The Inbox is the folder where all your new emails are stored. When you login, you will see how many messages you have in your Inbox and how many of them are new. Your emails are displayed one below the other, with the latest message first provided you have sorted your emails accordingly. Also shown is the sender of the email, the subject, the date it was sent and the size of the email.

If you want to make the best use of email make sure to check regularly for incoming mail and reply as soon as possible. The benefit of email is its ease of use in sending, receiving AND replying promptly.

Remote Access

You don't have to use the same computer to access your email, you can access it remotely from other computers as long as the computer has email software and you know the domain name of your mail server and your user details.

Reply Command

The email program has an automatic reply command, which will instruct it to prepare an email reply to go to the sender of the email. The program automatically insert the senders email address, and by default the same subject header. You have the option of including the original text in the message when you reply so that you can refer to points in it.

Bounced Email

Occasionally you will get email returned to you that has "bounced", this happens if you type the address incorrectly or the user has cancelled their account. It will be returned to you with an error message telling you why it couldn't be delivered like - "this message could not be delivered as there was no email account or mail box for this person". It is very important to get the address exactly right including all underscores and full stops.

Spam

Spam is unwanted or junk email that is sent to multiple recipients who have not specifically requested the email.

Never reply back to the sender of the Spam mail. Never even respond to the Spam email's instructions to reply with the word 'remove'. You may only be confirming that your email address is valid. It is best to delete the message and block the sender.

Select email(s) you identify as Spam and click on the 'This is Spam' button. This will block the sender and report the email as Spam to the Rediffmail anti-spam system.

The anti-spam system takes reports like these from you and other Rediffmail users and blocks incoming junk email (spam).

11.3. HTML

Hyper Text Markup Language, the coding language used to create hypertext documents for the World Wide Web. In HTML, a block of text can be surrounded with tags that indicate how it should appear (for example, in bold face or italics). Also, in HTML a word, a block of text, or an image can be linked to another file on the Web. HTML files are viewed with a World Wide Web browser.

> **Concept**
>
> *HTML is a language to create web pages.*

HTML is an abbreviation of "HyperText Mark-up Language" - which is already more than you need to know at this stage. However, for the sake of good order, let us explain in greater detail.

- **Hyper** is the opposite of linear. In the good old days - when a mouse was something the cat chased - computer programs ran linearly: when the program had executed one action it went to the next line and after that, the next line and so on. But HTML is different - you can go wherever you want and whenever you want. For example, it is not necessary to visit MSN.com before you visit HTML.net.
- **Text** is self-explanatory.
- **Mark-up** is what you do with the text. You are marking up the text the same way you do in a text editing program with headings, bullets and bold text and so on.
- **Language** is what HTML is. It uses many English words.

11.3.1. Simple rules of HTML

- Element names and attribute names are not case sensitive: attribute *values* **are**;
- Documents start with a <!doctype...> statement, followed by a header and a text body all enclosed in <html>...</html>
- The header is enclosed in <head>...</head>
- The text body is enclosed in <body>...</body>
- Comments are written as <!— A comment —>

To create HTML file Type in the following text in notepad:

```
<html>
<head>
<title>Title of page</title>
</head>
<body>
This is my first homepage. <b>This text is bold</b>
```

```
</body>
</html>
```

Save the file as "mypage.htm".

Start your Internet browser. Select "Open" (or "Open Page") in the File menu of your browser. A dialog box will appear. Select "Browse" (or "Choose File") and locate the HTML file you just created - "mypage.htm" - select it and click "Open". Now you should see an address in the dialog box, for example "C:\MyDocuments\mypage.htm". Click OK, and the browser will display the page.

The first tag in your HTML document is <html>. This tag tells your browser that this is the start of an HTML document. The last tag in your document is </html>. This tag tells your browser that this is the end of the HTML document.

The text between the <head> tag and the </head> tag is header information. Header information is not displayed in the browser window.

The text between the <title> tags is the title of your document. The title is displayed in your browser's caption.

The text between the <body> tags is the text that will be displayed in your browser.

The text between the <b> and </b> tags will be displayed in a bold font

11.3.2. HTML Tags

- HTML tags are used to mark-up HTML **elements**
- HTML tags are surrounded by the **two characters < and >**
- The surrounding characters are called **angle brackets**
- HTML tags normally **come in pairs** like <b> and </b>
- The first tag in a pair is the **start tag,** the second tag is the **end tag**
- The text between the start and end tags is the **element content**
- HTML tags are **not case sensitive,** <b> means the same as <B>

HTML Elements

Remember the HTML example :

```
<html>
<head>
<title>Title of page</title>
</head>
<body>
This is my first homepage. <b>This text is bold</b>
</body>
</html>
```

This is an HTML element:

```
<b>This text is bold</b>
```

The HTML element starts with a **start tag**: <b>

The **content** of the HTML element is: This text is bold

The HTML element ends with an **end tag**: </b>

The purpose of the <b> tag is to define an HTML element that should be displayed as bold.

This is also an HTML element:

```
<body>
This is my first homepage. <b>This text is bold</b>
</body>
```

This HTML element starts with the start tag <body>, and ends with the end tag </body>.

The purpose of the <body> tag is to define the HTML element that contains the body of the HTML document.

Tag Attributes

Tags can have attributes. Attributes can provide additional information about the HTML elements on your page.

This tag defines the body element of your HTML page: <body>. With an added bgcolor attribute, you can tell the browser that the background color of your page should be red, like this: <body bgcolor="red">.

This tag defines an HTML table: <table>. With an added border attribute, you can tell the browser that the table should have no borders: <table border="0">

Attributes always come in name/value pairs like this: name="value".

Attributes are always added to the start tag of an HTML element.

Quote Styles, "red" or 'red'?

Attribute values should always be enclosed in quotes. Double style quotes are the most common, but single style quotes are also allowed.

In some rare situations, like when the attribute value itself contains quotes, it is necessary to use single quotes:

name='John "ShotGun" Nelson'

The most important tags in HTML are tags that define headings, paragraphs and line breaks.

Headings

Headings are defined with the <h1> to <h6> tags. <h1> defines the largest heading. <h6> defines the smallest heading.

```
<h1>This is a heading</h1>
<h2>This is a heading</h2>
<h3>This is a heading</h3>
<h4    >This is a heading</h4>
<h5>This is a heading</h5>
<h6>This is a heading</h6>
```

HTML automatically adds an extra blank line before and after a heading.

Paragraphs

Paragraphs are defined with the <p> tag.

```
<p>This is a paragraph</p>
<p>This is another paragraph</p>
```

HTML automatically adds an extra blank line before and after a paragraph.

Line Breaks

The <br> tag is used when you want to end a line, but don't want to start a new paragraph. The <br> tag forces a line break wherever you place it.

```
<p>This <br> is a para<br>graph with line breaks</p>
```

The <br> tag is an empty tag. It has no closing tag.

Comments in HTML

The comment tag is used to insert a comment in the HTML source code. A comment will be ignored by the browser. You can use comments to explain your code, which can help you when you edit the source code at a later date.

```
<!— This is a comment —>
```

Note that you need an exclamation point after the opening bracket, but not before the closing bracket.

Basic Notes - Useful Tips

When you write HTML text, you can never be sure how the text is displayed in another browser. Some people have large computer displays, some have small. The text will be reformatted every time the user resizes his window. Never try to format the text in your editor by adding empty lines and spaces to the text.

HTML will truncate the spaces in your text. Any number of spaces count as one. Some extra information: In HTML a new line counts as one space.

Using empty paragraphs <p> to insert blank lines is a bad habit. Use the <br> tag instead. (But don't use the <br> tag to create lists. Wait until you have learned about HTML lists.)

You might have noticed that paragraphs can be written without the closing tag </p>. Don't rely on it. The next version of HTML will not allow you to skip ANY closing tags.

HTML automatically adds an extra blank line before and after some elements, like before and after a paragraph, and before and after a heading.

11.4. JAVA APPLETS

A web page designed using HTML is a static page. Thus, the information presented in a web page is constant. As with any visual medium, the dynamic behavior applied to a web page will make it more attractive. The dynamic behavior in web page can b in the form of:

- Scrolling text
- Messages displayed in the status bar
- Changing images

To be able to apply such behavior to a web page, one solution is to use a java applet.

An applet is a little application program. Prior to the World Wide Web, the built-in writing and drawing programs that came with Windows were sometimes called "applets." On the Web, using Java, the object-oriented programming language, an applet is a small program that can be sent along with a Web page to a user. Java applets can perform interactive animations, immediate calculations, or other simple tasks without having to send a user request back to the server.

Concept

An applet is a Java program used to create dynamic webpages on the browser with help of HTML.

A java applet is a program that can be embedded in a HTML page. Thus, we can write an Applet that will display scrolling text for example, and then embed it into the HTML page, giving it the ability to display scrolling text.

11.4.1. Java Applet Vs Applications

There are some of the differences between Applet and an application in Java:

- A Java application is a stand-alone program. This means that the application can be run by itself.
  A Java applet is a program that has to be run from within a web browser or a similar java enabled application such as an Applet Viewer.
- The Java applet is run from within a browser and so can make use of all the infrastructure that the browser can provide. For example, the event handling capability, user interface etc. These may also be specifically built into the Applet by the programmer.
- Java applet are included in HTML pages using the special tag <APPLET>. When the user loads a page containing a Java applet in his browser, the browser downloads the applet on to the local machine and executes the bytecode on the local system.
- Since the applet is executed on the local computer, the browser limit the actions that an applet can perform. This is done so that the applet does not cause any damage to the client system.

11.4.2. Restrictions on a Java Applet

Some of the most popular browser in the world today are the Netscape Navigator and the Internet Explorer.

The Following are some of the restrictions placed by these browser on the Java Applets:

- Applets are not allowed to read from or write to files on the local system.
- Applets can communicate only with the server on they are originally resident and with no other server either on the Interet or on the client's local network.
- Applets are not allowed to execute any application on the local computer neither are they allowed to load any DLLs.

11.4.3. The Structure of Java Applet

Consider the following code for Java Applet. The code given below is used to write a applet code which displays *Hello, World!.*

Example 11.1: Hello world

```
import java.applet.*;
import java.awt.*;

public class Our_Applet extends Applet
```

```
{
    String disp_str;
    public void init()
    {
        disp_str = "Hello, World !";
    }
    public void paint(Graphics g)
    {
        g.drawString(disp_str, 50, 25);
    }
}
```

The applet is a Java program that can be run only in a Browser or other similar Java enabled applications like an Appletviewer.

Embedding an applet in an HTML page is very simple and is done as shown below: Create an html file in note pad with following code:

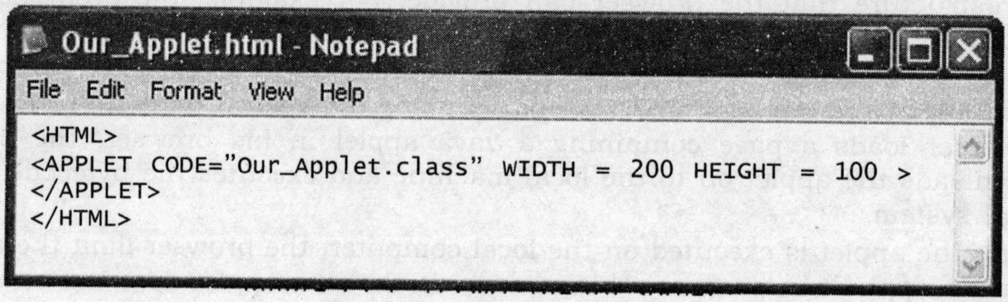

Fig. 11.1. HTML file for applet

```
<APPLET CODE="Our_Applet.class" WIDTH = 200 HEIGHT = 100 >
</APPLET>
```

The tag '<APPLET> </APPLET>' are used to signify that an Applet is being added to the HTML page. The attribute 'Code' is used to specify the name of the Applet's class file. The attributes 'Width' and 'Height' are used to define the applet's display area.

This HTML code should be saved as "Our_Apple.html".

To view the applet in an Appletviewer, HTML file created for the applet has to be run as shown below.

```
C:\JavaProgs> appletviewer <appletfilename>.html
```

For the applet created above , the applet program can be viewed as shown below

```
C:\JavaProgs> appletviewer Our_Applet.html
```

The output will appear as like the figure:

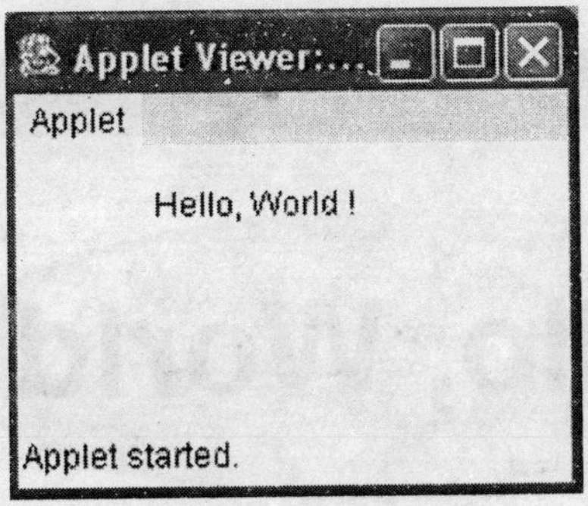

Fig. 11.2. Applet output

Applets can also be viewed with the help of Browser by opening the required HTML files in that browser.

To show the above example in more interesting graphical display, we have done some changes. This applet also does all of its drawing in the *paint( )* method. Applet shows the use of *java.awt.Font* and *java.awt.Color* objects.

Example 11.2: Fancy Hello World

```
import java.applet.*;
import java.awt.*;
public class Our_Applet extends Applet
{
    String st="Hello, World !";
    private Font font;
    public void init()
  {
    font=new Font("sansserif",Font.BOLD,48);
  }
    public void paint(Graphics g)
  {
    g.setColor(Color.pink);
    g.fillOval(10,10,330,100);

    //The red outline.
    //To simulate a 4-pixel wide line we draw four ovals.
        g.setColor(Color.red);
        g.drawOval(10,10,330,100);
        g.drawOval(9,9,332,102);
        g.drawOval(8,8,334,104);
        g.drawOval(7,7,336,106);

        //To display Text
        g.setColor(Color.black);
        g.setFont(font);
        g.drawString(st,40,75);
    }
  }
```

Output will looks like :

Fig. 11.3. Fancy Applet

The previous two applets have only displayed output. If an applet is to be interactive in any way, it has to receive and respond to user input. Example 11.3 shows a simple applet that lets the user do a freehand sketch with the mouse. See Figure 11.3.

The *mouseDown( )* and *mouseDrag( )* methods are called by the system when the user presses a mouse button and moves the mouse with the button down, respectively. This very simple applet draws lines directly in response to these events. It does not have a paint() method, which means that the user's scribble is lost any time that the applet is redrawn (for example, when a Web browser scrolls down a page and then scrolls back up again).

Note that both mouseDown() and mouseDrag() return true. This is to tell the system that they've handled the Event object that was passed to them, and that the event should not be processed any further.

Example 11.3: FreeHand Drawing

```
import java.applet.*;
import java.awt.*;
public class Scribble extends Applet
{
    private int last_x = 0, last_y = 0;

    public boolean mouseDown(Event e, int x, int y)
  {
    last_x = x; last_y = y;
    return true;
  }

  public boolean mouseDrag(Event e, int x, int y)
  {
    Graphics g = getGraphics(); // Get a Graphics to draw with.
    g.drawLine(last_x, last_y, x, y); // Draw a line from last point
    last_x = x; last_y = y;
    return true;
    }
}
```

Figure 11.4 shows the Output of the example 11.3 Applet

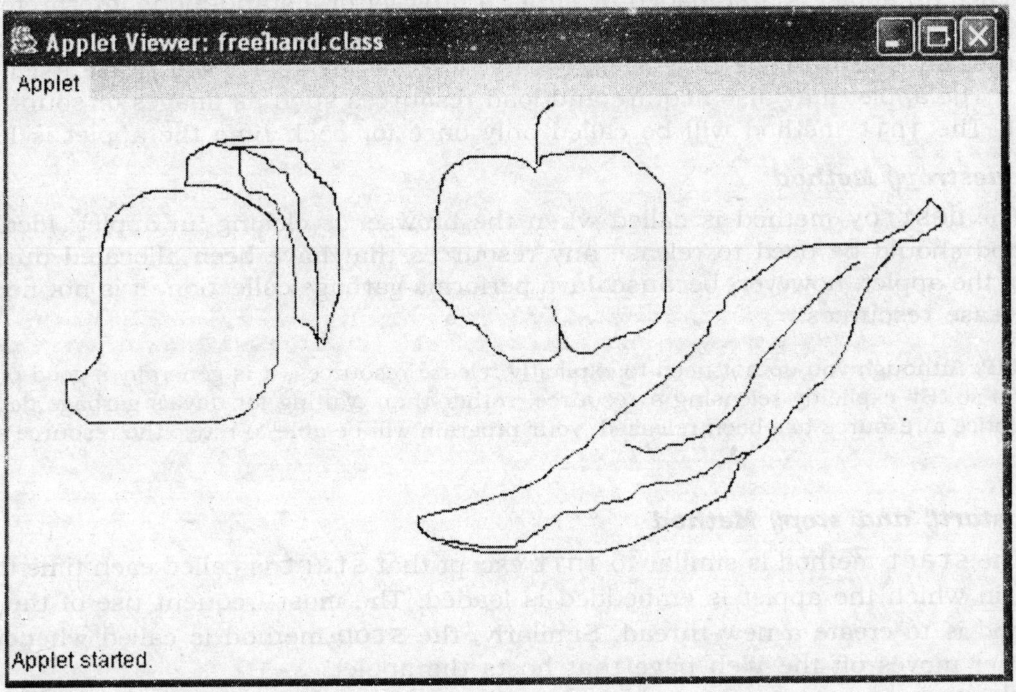

Fig. 11.4. Freehand Applet

11.5. THE APPLET CLASS

To create your own applets, you derive a class from Java's `Applet` class and then add the desired functionality. The `Applet` class contains methods for actions such as interacting with the browser environment in which it is running, for retrieving images, for playing sound files, and for resizing its host window. In this section you will read about the most frequently used methods of this important class.

11.5.1. The Life and Death of an Applet

On a typical Java-enhanced Web page, Java applets start and stop frequently as the user switches among pages. As pages are loaded into the browser, viewed and then left, applets are started and terminated in response to messages passed to the `Applet` class. The four most important methods in the life of an applet are `init`, `start`, `stop`, and `destroy`. They are described in Table 11.1.

Table 11.1. Important methods over the life of an applet

Method	Purpose
Init	Called when the applet is initially loaded.
Start	Called each time the user loads the host Web page.
Stop	Called each time the user leaves the host Web page.
Destroy	Called to release resources allocated by the applet.

The init() Method

When an applet is first loaded by either a browser or a stand-alone interpreter (such as is provided with Visual J++), the `init` method is called. This allows the applet to execute any startup code that is necessary. Usually an applet will resize itself at this point. The applet may also acquire and load resources such as images or sound at this point. The `init` method will be called only once for each time the applet is loaded.

The destroy() Method

The `destroy` method is called when the browser is closing an applet. Ideally this method should be used to release any resources that have been allocated during the life of the applet; however, because Java performs garbage collection, it is not necessary to release resources.

> **TIP:** Although you do not need to explicitly release resources, it is generally a good practice to do so. By explicitly releasing a resource, rather than waiting for Java's garbage detection to notice a resource has been released, your program will be able to reuse the resource sooner

The start() and stop() Method

The `start` method is similar to `init` except that `start` is called each time the Web page in which the applet is embedded is loaded. The most frequent use of the `start` method is to create a new thread. Similarly, the `stop` method is called whenever the browser moves off the Web page that hosts the applet.

The distinction between a method that gets called when an applet is first loaded and when its host page is first loaded might seem like a fine distinction, but understanding this distinction enables you to write more efficient applets. In Netscape Navigator 3 and Microsoft Internet Explorer 3, when a page containing a Java applet is first loaded, the `init` and `start` methods are called. Following a link from the page calls the applet's `stop` method. Backing up to the page calls the applet's `start` method, but not its `init` method.

In Netscape Navigator, reloading a page that contains an applet will stop and then start the applet. Performing the equivalent function in Microsoft Internet Explorer, however, causes the `stop`, `destroy`, `init`, and `start` methods all to be invoked in that order. For convenience, this information is summarized in Table 11.2.

Table 11.2. Applet method calls under Navigator and Internet Explorer

What	Navigator	Internet Explorer
Page first loaded	init + start	init + start
Forward then backward	stop + start	stop + start
Reload/Refresh	stop + start	stop + destroy + init + start

11.5.2. Example Applets

The best way to understand how the `init`, `start`, `stop`, and `destroy` methods interact is to build an applet to demonstrate their use. Example 11.2 shows a class that does exactly this. The class `EX04A` includes a private string member named `history` that is empty initially. As the various methods are called, text is added to this string to indicate the sequence in which the methods were called. Parentheses are placed around the start and stop messages to improve readability.

Example 11.4: EX04A.java.

```
import java.applet.*;
import java.awt.*;

public class EX04A extends Applet
{
    private String history = "";

    public EX04A()
    {
    }

    public void init()
    {
        history = history + "init ";
        resize(320, 240);
    }

 public void start()
 {
    history = history + "(start ";
 }
 public void stop()
 {
    history = history + "stop) ";
 }

 public void destroy()
 {
    history = history + "destroy ";
 }

 public void paint(Graphics g)
 {
    g.drawString(history, 10, 10);
 }
}
```

To execute the above example 11.4, create an HTML file of following code:

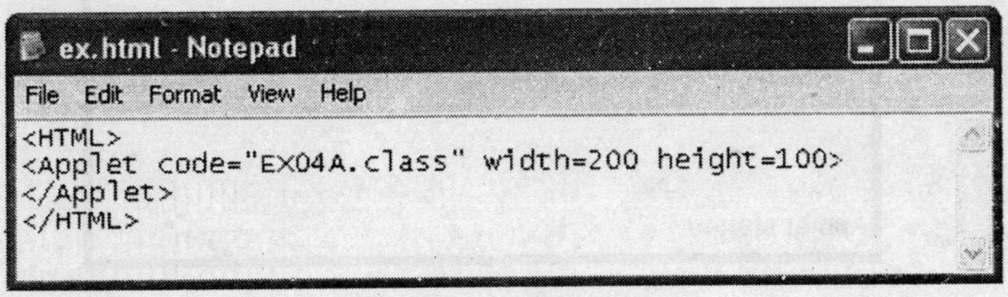


```
<HTML>
<Applet code="EX04A.class" width=200 height=100>
</Applet>
</HTML>
```

Fig. 11.5. HTML File

Even though the code in the destroy method of Example 11.4 will execute, the results of adding this text will not be apparent. Because the history string is displayed only by the paint method and paint is not called after destroy, you will not be able to see the destroy message.

To see the full message of above example write the following command in the command prompt and press restart option of Applet menu in appletviewer:

```
C:\JavaProgs>appletviewer ex.html
```

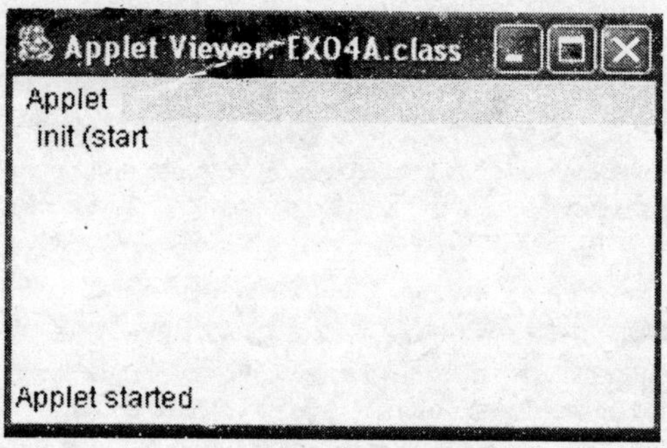

Fig. 11.6.

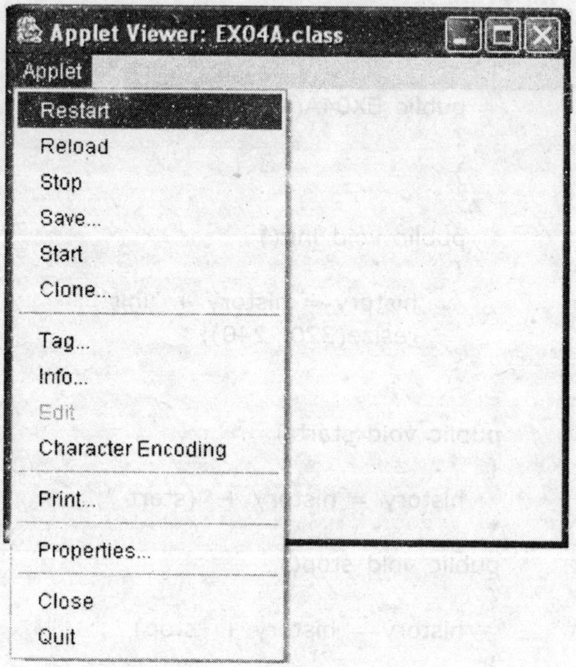

Fig. 11.7.

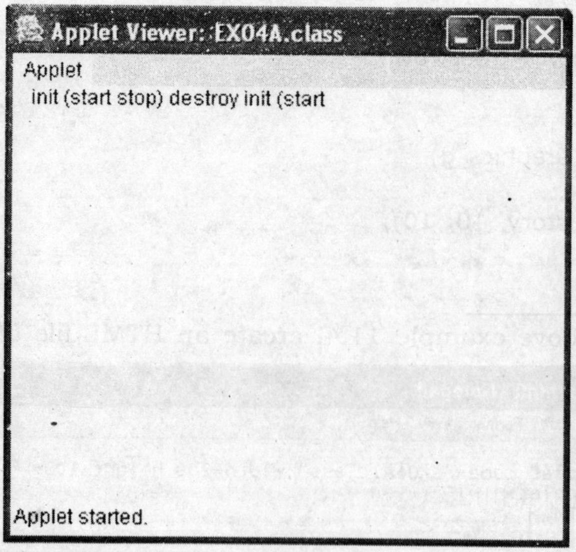

Fig. 11.8.

After the applet was loaded, the Restart button was pressed repeatedly. This had the effect of stopping, destroying and starting the applet each time, as you can see in Figure 11.8. Closing of appletviewer will stop the applet and then destroy it.

11.6. JSP

JavaServer Pages (JSP) is a technology based on the Java language and enables the development of dynamic web sites. JSP was developed by Sun Microsystems to allow server side development. JSP files are HTML files with special Tags containing Java source code that provide the dynamic content.

The following shows the Typical Web server, different clients connecting via the Internet to a Web server. In this example, the Web server is running on Unix and is the very popular Apache Web server.

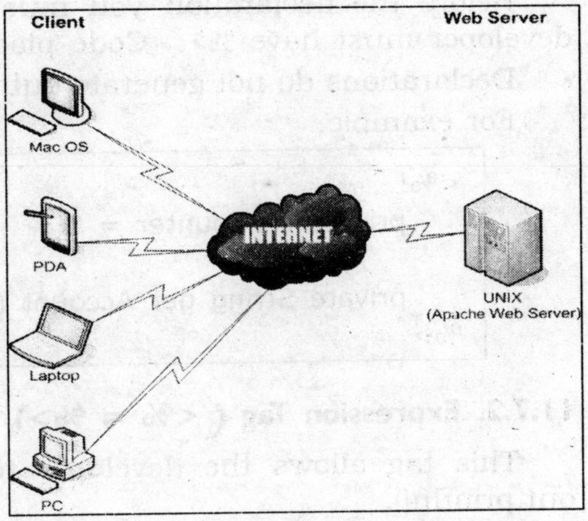

Fig. 11.9.

First static web pages were displayed. Typically these were people's first experience with making web pages so consisted of My Home Page sites and company marketing information. Afterwards Perl and C were languages used on the web server to provide dynamic content. Soon most languages including Visualbasic,Delphi,C and Java could be used to write applications that provided dynamic content using data from text files or database requests. These were known as CGI server side applications. ASP was developed by Microsoft to allow HTML developers to easily provide dynamic content supported as standard by Microsoft's free Web Server, Internet Information Server (IIS). JSP is the equivalent from Sun Microsystems, a comparison of ASP and JSP will be presented in the following section.

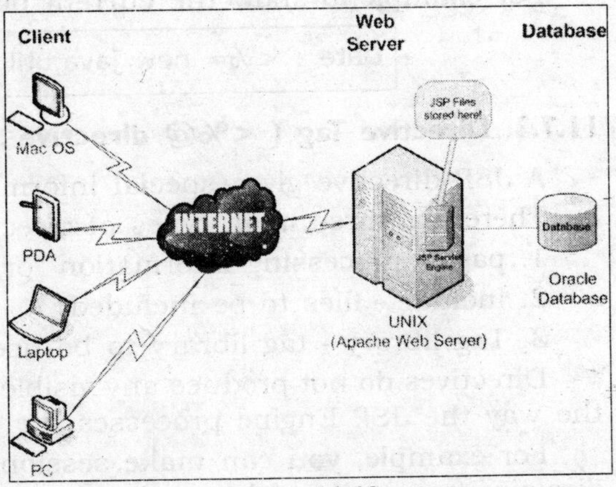

Fig. 11.10.

The following diagram shows a web server that supports JSP files. Notice that the web server also is connected to a database.

JSP source code runs on the web server in the JSP Servlet Engine. The JSP Servlet engine dynamically generates the HTML and sends the HTML output to the client?s web browser.

11.7. JSP TAGS

There are five main tags:

1. Declaration tag
2. Expression tag
3. Directive tag

4. Scriptlet tag
5. Action tag

11.7.1. Declaration Tag (<%! %>)

This tag allows the developer to declare variables or methods.

Before the declaration you must have <%! , at the end of the declaration, the developer must have %> . Code placed in this tag must end in a semicolon (;).

Declarations do not generate output so are used with JSP expressions or scriptlets. For example,

```
<%!
     private int counter = 0 ;

     private String get Account ( int accountNo) ;
%>
```

11.7.2. Expression Tag (<% = %>)

This tag allows the developer to embed any Java expression and is short for out.println().

A semicolon (;) does not appear at the end of the code inside the tag.

For example,to show the current date and time.

```
Date : <%= new java.util.Date() %>
```

11.7.3. Directive Tag (<%@ directive ... %>)

A JSP directive gives special information about the page to the JSP Engine.

There are three main types of directives:

1. page - processing information for this page.
2. Include - files to be included.
3. Tag library - tag library to be used in this page.

Directives do not produce any visible output when the page is requested but change the way the JSP Engine processes the page.

For example, you can make session data unavailable to a page by setting a page directive (session) to false.

1. Page Directive

This directive has 11 optional attributes that provide the JSP Engine with special processing information. The following table lists the 11 different attributes with a brief description:

Attribute	Description	Example
language	Which language the file uses.	<%@ page language = "java" %>
extends	Superclass used by the JSP engine for the translated Servlet.	<%@ page extends = "com.taglib..%>
import	Import all the classes in a java package into the current JSP page. This allows the JSP page to use other java classes.	<%@ page import = "java.util.*"%>

Attribute	Description	Example
session	Does the page make use of sessions. By default all JSP pages have session data available. There are performance benefits to switching session to false.	`Default is set to true.`
Buffer	Controls the use of buffered output for a JSP page. Default is 8 kb	`<%@ page buffer = "none" %>`
autoFlush	Flush output buffer when full.	`<%@ page autoFlush = "true" %>`
isThreadSafe	Can the generated Servlet deal with multiple requests? If true a new thread is started so requests are handled simultaneously.	
info	Developer uses info attribute to add information/document for a page. Typically used to add author,version,copyright and date info.	`<%@ page info = "visualbuilder. comtest page,copyright 2001. " %>`
errorPage	Different page to deal with errors. Must be URL to error page.	`<%@ page errorPage = "/error/error.jsp" %>`
IsErrorPage	This flag is set to true to make a JSP page a special Error Page. This page has access to the implicit object exception (see later).	
contentType	Set the mime type and character set of the JSP.	

2. *Include Directive*

Allows a JSP developer to include contents of a file inside another. Typically include files are used for navigation, tables, headers and footers that are common to multiple pages.

Two examples of using include files:

This includes the html from privacy.html found in the include directory into the current jsp page.

```
<%@ include file = "include/privacy.html" %>
```

or to include a naviagation menu (jsp file) found in the current directory.

```
<%@ include file = "navigation.jsp" %>
```

Include files are discussed in more detail in the later sections.

3. *Tag Lib Directive*

A tag lib is a collection of custom tags that can be used by the page.

```
<%@ taglib uri = "tag library URI" prefix = "tag Prefix" %>
```

Custom tags were introduced in JSP 1.1 and allow JSP developers to hide complex server side code from web designers.

11.7.4. Scriptlet Tag (<% ... %>)

Between <% and %> tags,any valid Java code is called a Scriptlet. This code can access any variable or bean declared.

For example, to print a variable.

```
<%
    String username = "Superman";
    out.println ( username ) ;
%>
```

11.7.5. Action Tag

There are three main roles of action tags :

1. enable the use of server side Javabeans
2. transfer control between pages
3. browser independent support for applets.

A Javabean is a special type of class that has a number of methods. The JSP page can call these methods so can leave most of the code in these Javabeans. For example, if you wanted to make a feedback form that automatically sent out an email. By having a JSP page with a form, when the visitor presses the submit button this sends the details to a Javabean that sends out the email. This way there would be no code in the JSP page dealing with sending emails (JavaMail API) and your Javabean could be used in another page (promoting reuse).

To use a Javabean in a JSP page use the following syntax:

```
<jsp : usebean id = " ...." scope = "application" class = "com..." />
```

The following is a list of Javabean scopes:

- page-valid until page completes.
- request-bean instance lasts for the client request.
- session-bean lasts for the client session.
- application-bean instance created and lasts until application ends.

11.8. CREATING A FORM.

Here we show an example to create and process an html form.

Write following code and save as a file named: myform.jsp and open the file via browser.

```
<html>
<head>
<!— Example4 —>
<title>My Form</title>
</head>
<body>
<form action="myformconfirm.jsp" method="post">
Enter in a website name:<br>
<input type="text" name="website"><br>
<input type="submit" name="submit">
</form>
</body>
</html>
```

Processing a Form

Here we show all steps, how to process the html form you just created.

- Copy the code below and place in a file named: myformconfirm.jsp
- Go to myform.jsp
- Fill in some details and submit the form

You should see the results of your submission

```
<html>
<head>
<!— Example4 —>
<title>My Form</title>
</head>
<body>
<font size=3>
Your info has been received:
<br><br>
<%
String sName = request.getParameter("website");
out.print(sName);
%>
</font>
</body>
</html>
```

SUMMARY

In this chapter we have explained the concept and style to design web pages using HTML and Java Applets.

- HTML is a language to create webpages.
- Java Applet is a Java program used to create dynamic web pages on the browser with the help of HTML.
- At last we have given the functionality of Applet Class of Java. We have provide most frequently used methods of this important class.

SOLVED PROBLEMS

Q.11.1. What is the main difference between "APPLET" and "APPLICATION"

Ans. An application is a standalone Java program characterized by the public static void main (String[] args){} method.In contrast,an applet is a java program launched through a browser.The main method is the entry point for a Java application program whereas init() method of the applet is its entry point for the program.

Q.11.2. How will you initialize an Applet?

Ans. Using init() method.

Q.11.3. What is the order of method invocation in an Applet?

Ans. The following is order:

1. init()
2. start()
3. paint()
4. stop()
5. destroy()

Q.11.4. When is update method called?

Ans. update method is called every time whenever we call a repaint() method.

Q.11.5. What are the guidelines for HTML formatted email?

Ans. The first (un-helpful) response is that email is a plain text medium and sending other MIME types is bound to be a problem for *some* users. One of the great advantages of email is that it is lightweight and usually takes a fraction of a second to download. If you include surplus HTML markup and perhaps image references you will significantly impair the user experience and turn subscribers away from your service, either by deleting the mail without reading or un-subscribing

Q.11.6. What's the difference between reloading and restarting an applet?

Ans. Reloading clears all static variables and starts over again by constructing a new instance of the applet and calling its init() method. Depending on the platform reload may or may not reload the byte code from a .class file on the disk or on the network. Restarting an applet just calls its start() method. It does not call init(), the constructor, or reload the byte codes.

Q.11.7. Assume that each letter occupies a box fifteen pixels wide by fifteen pixels high. Write an applet which takes a single letter as a parameter and fills the available space with that letter.

Ans.

```
import java.applet.Applet;
import java.awt.Graphics;

public class letterFillApplet extends Applet {
   String theLetter;

   public void init() {

   String s = getParameter("Letter");
   theLetter = s.substring(0, 1);

 }

      public void paint(Graphics g) {

      for (int i = 0; i < size().width; i += 15) {
      for (int j = 0; j < size().height; j += 15) {
      g.drawString(theLetter, i, j);
      }
   }

 }

}
```

Q.11.8. Create an applet for the PoetryApplet so that it includes a numlines parameter that will set the size of the poem array at runtime.

Ans.

```
    import java.applet.Applet;
    import java.awt.Graphics;

public class PoetryApplet extends Applet {

    String[] poem;
    int numlines;

    public void init() {

    int i;
    String nextline;

  try {
    String s = getParameter("numlines");
    if (s == null) numlines = 101;
    else numlines = Integer.parseInt(s);
  }
    catch (NumberFormatException e) {
    numlines = 101;
  }

  poem = new String[numlines];

  for (i = 1; i <= numlines; i++) {
    nextline = getParameter("Line" + i);
    if (nextline == null) break;
    poem[i] = nextline;
  }
    numlines = i-1;

  }

    public void paint(Graphics g) {

    int y = 15;

    for (int i=1; i <= numlines; i++) {
    g.drawString(poem[i], 5, y);
    y += 15;
    }

  }

}
```

Q.11.9. Change the PoetryApplet to use a Vector rather than an array so that it will truly handle an arbitrary number of parameters.

Ans.

```
import java.applet.Applet;
import java.awt.Graphics;
import java.util.Vector;
import java.util.Enumeration;

public class PoetryApplet extends Applet {

Vector poem;

 public void init() {

 int i = 1;
 String nextline;
poem = new Vector();

 while ((nextline = getParameter("Line" + i++)) != null) {
    poem.addElement(nextline);
}

}

public void paint(Graphics g) {

int y = 15;

Enumeration e = poem.elements();
while (e.hasMoreElements()) {
g.drawString((String) e.nextElement(), 5, y);
y += 15;
}

}

}
```

Q.11.10. Create a WrapText applet so that it optionally accepts FONT, SIZE and STYLE PARAMs which set the font face, point size and style in which the String is drawn. Be sure to gracefully handle the case when the APPLET tag does not include any or all of these PARAMs or includes invalid values (e.g. STYLE="UNDERLINE").

Ans.

```
import java.applet.Applet;
import java.awt.Graphics;
import java.awt.FontMetrics;
import java.util.StringTokenizer;
import java.awt.Font;
```

```
public class WrapTextApplet extends Applet
{
 String input_from_page;
 int margin = 5;
 Font theFont;
 public void init() {

 String s;
 int fontsize = 12;
 String fontname = "Courier";
 int fontstyle = Font.PLAIN;

 input_from_page = getParameter("Text");
 if ((s = getParameter("Size")) != null) {
 fontsize = Integer.parseInt(s);
 }
 if ((s = getParameter("Style")) != null) {
 if (s.equalsIgnoreCase("Italic")) fontstyle = Font.ITALIC;
 if (s.equalsIgnoreCase("Bold")) fontstyle = Font.BOLD;
 }
 if ((s = getParameter("Font")) != null) {
 fontname = s;
 }
 theFont = new Font(fontname, fontstyle, fontsize);
 }
 public void paint(Graphics g) {
 int i = 0;
 int linewidth = 0;
 StringBuffer sb = new StringBuffer();
 g.setFont(theFont);
    FontMetrics fm = g.getFontMetrics();
    StringTokenizer st = new StringTokenizer(input_from_page);
    while (st.hasMoreTokens()) {
     String nextword = st.nextToken();
     if (fm.stringWidth(sb.toString() + nextword) < size().width) {
    sb.append(nextword);
     sb.append(' ');
     }
     else if (sb.length() == 0) {
     g.drawString(nextword, margin, ++i*fm.getHeight());
     }
     else {
 g.drawString(sb.toString(), margin, ++i*fm.getHeight());
 sb = new StringBuffer(nextword + " ");
 }

 }
 if (sb.length() > 0) {
```

```
g.drawString(sb.toString(), margin, ++i*fm.getHeight());
}
}

}
```

EXERCISES

Q.11.1. What is the use of HTML?

Q.11.2. List out the Basic rules of HTML?

Q.11.3. What is the requirement of Java applets?

Q.11.4. Differentiate between Applet and application.

Q.11.5. Explain the life cycle of Java Applets

Q.11.6. What is the role of Applet class?

Q.11.7. Can we use the constructors in Applets?

Q.11.8. Create an applet in which one string moves from left to right.

Q.11.9. Creat an applet to display the shape of triangle, Rectangle, circle with different colors.

Q.11.10. Create an applet to display following figure.

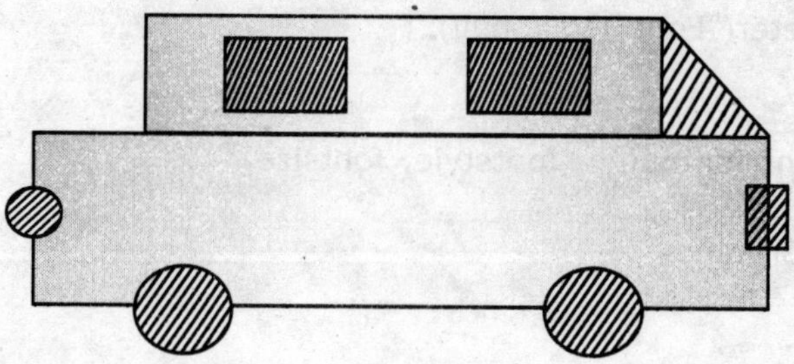

Q.11.11. Explain the following tags:

(*a*) <head> (*b*) <h1>, <h2>

(*c*) <body> (*d*) <b>

(*e*) <title> (*f*) <p>

(*g*) <applet>

□□□

Appendix A

PLACEMENT QUESTIONS

Q.1. Describe the principles of OOPS.

Ans. There are three main principals of oops which are called Polymorphism, Inheritance and Encapsulation.

Q.2. Explain the Encapsulation principle.

Ans. Encapsulation is a process of binding or wrapping the data and the codes that operates on the data into a single entity. This keeps the data safe from outside interface and misuse. One way to think about encapsulation is as a protective wrapper that prevents code and data from being arbitrarily accessed by other code defined outside the wrapper.

Q.3. Explain the Inheritance principle.

Ans. Inheritance is the process by which one object acquires the properties of another object.

Q.4. Explain the Polymorphism principle.

Ans. The meaning of Polymorphism is something like one name many forms. Polymorphism enables one entity to be used as as general category for different types of actions. The specific action is determined by the exact nature of the situation. The concept of polymorphism can be explained as "one interface, multiple methods".

Q.5. Explain the different forms of Polymorphism.

Ans. From a practical programming viewpoint, polymorphism exists in three distinct forms in Java:

Method overloading

Method overriding through inheritance

Method overriding through the Java interface

Q.6. What are Access Specifiers available in Java?

Ans. ccess specifiers are keywords that determines the type of access to the member of a class. These are:

Public

Protected

Private

Defaults

Q.7. Describe the wrapper classes in Java.

Ans. Wrapper class is wrapper around a primitive data type. An instance of a wrapper class contains, or wraps, a primitive value of the corresponding type.

Following table lists the primitive types and the corresponding wrapper classes:

Primitive Wrapper
boolean java.lang.Boolean
byte java.lang.Byte
char java.lang.Character
double java.lang.Double
float java.lang.Float
int java.lang.Integer
long java.lang.Long
short java.lang.Short
void java.lang.Void

Q.8. Is Iterator a Class or Interface? What is its use?

Ans. Iterator is an interface which is used to step through the elements of a Collection.

Q.9. What is similarities/difference between an Abstract class and Interface?

Ans. **Differences are as follows:** Interfaces provide a form of multiple inheritance. A class can extend only one other class. Interfaces are limited to public methods and constants with no implementation. Abstract classes can have a partial implementation, protected parts, static methods, etc.

A Class may implement several interfaces. But in case of abstract class, a class may extend only one abstract class. Interfaces are slow as it requires extra indirection to to find corresponding method in in the actual class. Abstract classes are fast.

Similarities: Neither Abstract classes or Interface can be instantiated.

Q.10. If a class is located in a package, what do you need to change in the OS environment to be able to use it?

Q.11. You need to add a directory or a jar file that contains the package directories to the CLASSPATH environment variable. Let's say a class Employee belongs to a package com.xyz.hr; and is located in the file c:\dev\com\xyz\hr\Employee.java. In this case, you'd need to add c:\dev to the variable CLASSPATH. If this class contains the method main(), you could test it from a command prompt window as follows:

Ans. c:\>java com.xyz.hr.Employee

Q.12. How many methods in the Serializable interface?

Ans. There is no method in the Serializable interface. The Serializable interface acts as a marker, telling the object serialization tools that your class is serializable.

Q.13. What is the difference between Serializalble and Externalizable interface?

Ans. When you use Serializable interface, your class is serialized automatically by default. But you can override writeObject() and readObject() two methods to control more complex object serailization process. When you use Externalizable interface, you have a complete control over your class's serialization process.

Q.14. What is a transient variable in Java?

Ans. A transient variable is a variable that may not be serialized. If you don't want some field to be serialized, you can mark that field transient or static.

Q.15. What is Java?

Ans. Java is an object-oriented programming language developed initially by James Gosling and colleagues at Sun Microsystems. The language, initially called Oak (named after the oak trees outside Gosling's office), was intended to replace C++, although the feature set better resembles that of Objective C. Java should not be confused with JavaScript, which shares only the name and a similar C-like syntax. Sun Microsystems currently maintains and updates Java regularly.

Q.16. Can you have virtual functions in Java?

Ans. Yes, all functions in Java are virtual by default. This is actually a pseudo trick question because the word "virtual" is not part of the naming convention in Java (as it is in C++, C-sharp and VB.NET), so this would be a foreign concept for someone who has only coded in Java. Virtual functions or virtual methods are functions or methods that will be redefined in derived classes.

Q.17. What is more advisable to create a thread, by implementing a Runnable interface or by extending Thread class?

Ans. Strategically speaking, threads created by implementing Runnable interface are more advisable. If you create a thread by extending a thread class, you cannot extend any other class. If you create a thread by implementing Runnable interface, you save a space for your class to extend another class now or in future.

Q.18. What is NullPointerException and how to handle it?

Q.19. When an object is not initialized, the default value is null. When the following things happen, the NullPointerException is thrown:

- Calling the instance method of a null object.
- Accessing or modifying the field of a null object.
- Taking the length of a null as if it were an array.
- Accessing or modifying the slots of null as if it were an array.
- Throwing null as if it were a Throwable value.

The NullPointerException is a runtime exception. The best practice is to catch such exception even if it is not required by language design.

Q.20. What's the difference between an interface and an abstract class?

Ans. An abstract class may contain code in method bodies, which is not allowed in an interface. With abstract classes, you have to inherit your class from it and Java does not allow multiple inheritance. On the other hand, you can implement multiple interfaces in your class.

Q.21. What is synchronization and why is it important?

Ans. With respect to multithreading, synchronization is the capability to control the access of multiple threads to shared resources. Without synchronization, it is possible for one thread to modify a shared object while another thread is in the process of using or updating that object's value. This often causes dirty data and leads to significant errors.

Q.22. What are three ways in which a thread can enter the waiting state?

Ans. A thread can enter the waiting state by invoking its sleep() method, by blocking on IO, by unsuccessfully attempting to acquire an object's lock, or by invoking an object's wait() method. It can also enter the waiting state by invoking its (deprecated) suspend() method.

Q.23. What would you use to compare two String variables - the operator == or the method equals()?

Ans. I'd use the method equals() to compare the values of the Strings and the == to check if two variables point at the same instance of a String object.

Q.24. What is thread?

Ans. A thread is an independent path of execution in a system.

Q.25. What is multi-threading?

Ans. Multi-threading means various threads that run in a system.

Q.26. How does multi-threading take place on a computer with a single CPU?

Ans. The operating system's task scheduler allocates execution time to multiple tasks. By quickly switching between executing tasks, it creates the impression that tasks execute sequentially.

Q.27. How to create a thread in a program?

Ans. You have two ways to do so. First, making your class "extends" Thread class. Second, making your class "implements" Runnable interface. Put jobs in a run() method and call start() method to start the thread.

Q.28. What invokes a thread's run() method?

Ans. After a thread is started, via its start() method of the Thread class, the JVM invokes the thread's run() method when the thread is initially executed.

Q.29. What is the difference between yielding and sleeping?

Ans. When a task invokes its yield() method, it returns to the ready state. When a task invokes its sleep() method, it returns to the waiting state.

Q.30. What is the difference between Process and Thread?

Ans. A process can contain multiple threads. In most multithreading operating systems, a process gets its own memory address space; a thread doesn't. Threads typically share the heap belonging to their parent process. For instance, a JVM runs in a single process in the host O/S. Threads in the JVM share the heap belonging to that process; that's why several threads may access the same object. Typically, even though they share a common heap, threads have their own stack space. This is how one thread's invocation of a method is kept separate from another's. This is all a gross oversimplification, but it's accurate enough at a high level. Lots of details differ between operating systems. Process vs. Thread A program vs. similar to a sequential program an run on its own vs. Cannot run on its own Unit of allocation vs. Unit of execution Have its own memory space vs. Share with others Each process has one or more threads vs. Each thread belongs to one process Expensive, need to context switch vs. Cheap, can use process memory and may not need to context switch More secure. One process cannot corrupt another process vs. Less secure. A thread can write the memory used by another thread

Q.31. Can an inner class declared inside of a method access local variables of this method?

Ans. It's possible if these variables are final.

Q.32. What is the Vector class?

Ans. The Vector class provides the capability to implement a growable array of objects.

Q.33. What modifiers may be used with an inner class that is a member of an outer class?

Ans. A (non-local) inner class may be declared as public, protected, private, static, final, or abstract.

Q.34. If a method is declared as protected, where may the method be accessed?

Ans. A protected method may only be accessed by classes or interfaces of the same package or by subclasses of the class in which it is declared.

Q.35. What's the main difference between a Vector and an ArrayList?

Ans. Java Vector class is internally synchronized and ArrayList is not.

Q.36. What are wrapped classes?

Ans. Wrapped classes are classes that allow primitive types to be accessed as objects.

Q.37. Does garbage collection guarantee that a program will not run out of memory?

Ans. No, it doesn't. It is possible for programs to use up memory resources faster than they are garbage collected. It is also possible for programs to create objects that are not subject to garbage collection.

Q.38. What is the difference between preemptive scheduling and time slicing?

Ans. Under preemptive scheduling, the highest priority task executes until it enters the waiting or dead states or a higher priority task comes into existence. Under time slicing, a task executes for a predefined slice of time and then reenters the pool of ready tasks. The scheduler then determines which task should execute next, based on priority and other factors.

Q.39. Name Component subclasses that support painting ?

Ans. The Canvas, Frame, Panel, and Applet classes support painting.

Q.40. What is a native method?

Ans. A native method is a method that is implemented in a language other than Java.

Q.41. How can you write a loop indefinitely?

Ans. for(;;)—for loop; while(true)—always true, etc.

Q.42. Can an anonymous class be declared as implementing an interface and extending a class?

Ans. An anonymous class may implement an interface or extend a superclass, but may not be declared to do both.

Q.43. What is the purpose of finalization?

Ans. The purpose of finalization is to give an unreachable object the opportunity to perform any cleanup processing before the object is garbage collected.

Q.44. How does Java handle integer overflows and underflows?

Ans. It uses low order bytes of the result that can fit into the size of the type allowed by the operation.

Q.45. What is the difference between the Boolean & operator and the && operator?

Ans. If an expression involving the Boolean & operator is evaluated, both operands are evaluated. Then the & operator is applied to the operand. When an expression involving the && operator is evaluated, the first operand is evaluated. If the first operand returns a value of true then the second operand is evaluated. The && operator is then applied to the first and second operands. If the first operand evaluates to false, the evaluation of the second operand is skipped.

Operator & has no chance to skip both sides evaluation and && operator does. If asked why, give details as above.

Q.46. How can a subclass call a method or a constructor defined in a superclass?

Ans. Use the following syntax: super.myMethod(); To call a constructor of the superclass, just write super(); in the first line of the subclass's constructor.

Q.47. What is the purpose of the System class?

Ans. The purpose of the System class is to provide access to system resources.

Q.48. What is the purpose of the finally clause of a try-catch-finally statement?

Ans. The finally clause is used to provide the capability to execute code no matter whether or not an exception is thrown or caught.

Q.49. What is an abstract method?

Ans. An abstract method is a method whose implementation is deferred to a subclass. Or, a method that has no implementation.

Q.50. What is the difference between interface and abstract class?

Ans. Interface contains methods that must be abstract; abstract class may contain concrete methods. interface contains variables that must be static and final; abstract class may contain non-final and final variables. members in an interface are public by default, abstract class may contain non-public members. interface is used to "implements"; whereas abstract class is used to "extends". interface can be used to achieve multiple inheritance; abstract class can be used as a single inheritance. interface can "extends" another interface, abstract class can "extends" another class and "implements" multiple interfaces. interface is absolutely abstract; abstract class can be invoked if a main() exists. interface is more flexible than abstract class because one class can only "extends" one super class, but "implements" multiple interfaces. If given a choice, use interface instead of abstract class.

Q.51. What is a static method?

Ans. A static method is a method that belongs to the class rather than any object of the class and doesn't apply to an object or even require that any objects of the class have been instantiated.

Q.52. What is a protected method?

Ans. A protected method is a method that can be accessed by any method in its package and inherited by any subclass of its class.

Q.53. What is the difference between a static and a non-static inner class?

Ans. A non-static inner class may have object instances that are associated with instances of the class's outer class. A static inner class does not have any object instances.

Q.54. What classes of exceptions may be caught by a catch clause?

Ans. A catch clause can catch any exception that may be assigned to the Throwable type. This includes the Error and Exception types.

Q.55. What is the difference between throw and throws keywords?

Ans. The throw keyword denotes a statement that causes an exception to be initiated. It takes the Exception object to be thrown as argument. The exception will be caught by an immediately encompassing try-catch construction or propagated further up the calling hierarchy.

The throws keyword is a modifier of a method that designates that exceptions may come out of the method, either by virtue of the method throwing the exception itself or because it fails to catch such exceptions that a method it calls may throw.

Q.56. Does a class inherit the constructors of its super class?

Ans. A class does not inherit constructors from any of its superclasses.

Q.57. Name primitive Java types.

Ans. The primitive types are byte, char, short, int, long, float, double, and Boolean.

Q.58. What is the difference between static and non-static variables?

Ans. A static variable is associated with the class as a whole rather than with specific instances of a class. Non-static variables take on unique values with each object instance.

Q.59. What is the difference between the paint() and repaint() methods?

Ans. The paint() method supports painting via a Graphics object. The repaint() method is used to cause paint() to be invoked by the AWT painting thread.

Q.60. What is the purpose of the File class?

Ans. The File class is used to create objects that provide access to the files and directories of a local file system.

Q.61. What restrictions are placed on method overriding?

Ans. Overridden methods must have the same name, argument list, and return type. The overriding method may not limit the access of the method it overrides. The overriding method may not throw any exceptions that may not be thrown by the overridden method.

Q.62. What is casting?

Ans. There are two types of casting, casting between primitive numeric types and casting between object references. Casting between numeric types is used to convert larger values, such as double values, to smaller values, such as byte values. Casting between object references is used to refer to an object by a compatible class, interface, or array type reference.

Q.63. How are this() and super() used with constructors?

Ans. this() is used to invoke a constructor of the same class. super() is used to invoke a superclass constructor.

Q.64. How is it possible for two String objects with identical values not to be equal under the == operator? How are this() and super() used with constructors?

Ans. The == operator compares two objects to determine if they are the same objects in memory. It is possible for two String objects to have the same value, but located in different areas of memory.

Q.65. How can you force garbage collection?

Ans. You can't force GC, but could request it by calling System.gc(). JVM does not guarantee that GC will be started immediately.

Q.66. What are the Object and Class classes used for?

Ans. The Object class is the highest-level class in the Java class hierarchy. The Class class is used to represent the classes and interfaces that are loaded by a Java program.

Q.67. What is Serialization and deserialization ?

Ans. Serialization is the process of writing the state of an object to a byte stream. Deserialization is the process of restoring these objects.

Q.68. Explain the usage of Java packages.

Ans. This is a way to organize files when a project consists of multiple modules. It also helps resolve naming conflicts when different packages have classes with the same names. Packages access level also allows you to protect data from being used by the non-authorized classes.

Q.69. Is Java a super set of JavaScript?

Ans. No. They are completely different. Some syntax may be similar.

Q.70. How the object oriented approach helps us keep complexity of software development under control?

Ans. We can discuss such issue from the following aspects:

Objects allow procedures to be encapsulated with their data to reduce potential interference.

Inheritance allows well-tested procedures to be reused and enables changes to make once and have effect in all relevant places.

The well-defined separations of interface and implementation allow constraints to be imposed on inheriting classes while still allowing the flexibility of overriding and overloading.

Q.71. What is polymorphism?

Ans. Polymorphism means "having many forms". It allows methods (may be variables) to be written that needn't be concerned about the specifics of the objects they will be applied to. That is, the method can be specified at a higher level of abstraction and can be counted on to work even on objects of un-conceived classes.

Q.72. What is the benefit of subclass?

Ans. **Generally:** The sub class inherits all the public methods and the implementation.

The sub class inherits all the protected methods and their implementation.

The sub class inherits all the default(non-access modifier) methods and their implementation.

The sub class also inherits all the public, protected and default member variables from the super class.

The constructors are not part of this inheritance model.

Q.73. In System.out.println(),what is System,out and println,pls explain?

Ans. System is a predefined final class,out is a PrintStream object acting as a field member and println is a built-in overloaded method in the out object.

Q.74. Can you write a Java class that could be used both as an applet as well as an application?

Ans. Yes. Add a main() method to the applet.

Q.75. What is the output of x > y? a:b = p*q when x=1,y=2,p=3,q=4?

Ans. When this kind of question has been asked, find the problems you think is necessary to ask back before you give an answer. Ask if variables a and b have been declared or initialized. If the answer is yes. You can say that the syntax is wrong. If the statement is rewritten as: x

Q.76. Why Java does not support pointers?

Ans. Because pointers are unsafe. Java uses reference types to hide pointers and programmers feel easier to deal with reference types without pointers. This is why Java and C-sharp shine.

Q.77. Can you declare a class as private?

Ans. Yes, we can declare a private class as an inner class.

What is the difference between shallow copy and deep copy?

Shallow copy shares the same reference with the original object like cloning, whereas the deep copy get a duplicate instance of the original object. If the shallow copy has been changed, the original object will be reflected and vice versa.

Q.78. Why is multiple inheritance not possible in Java?

Ans. It depends on how you understand "inheritance". Java can only "extends" one super class, but can "implements" many interfaces; that doesn't mean the multiple inheritance is not possible. You may use interfaces to make inheritance work for you. Or you may need to work around. For example, if you cannot get a feature from a class because your class has a super class already, you may get that class's feature by declaring it as a member field or getting an instance of that class. So the answer is that multiple inheritance in Java is possible.

Q.79. What's the difference between constructors and other methods?

Ans. Constructors must have the same name as the class and can not return a value. They are only called once while regular methods could be called many times.

Q.80. What are the drawbacks of inheritance?

Ans. Since inheritance inherits everything from the super class and interface, it may make the subclass too clustering and sometimes error-prone when dynamic overriding or dynamic overloading in some situation. In addition, the inheritance may make peers hardly understand your code if they don't know how your super-class acts and add learning curve to the process of development.

Usually, when you want to use a functionality of a class, you may use subclass to inherit such function or use an instance of this class in your class. Which is better, depends on your specification.

Q.81. Is there any other way that you can achieve inheritance in Java?

Ans. There are a couple of ways. As you know, the straight way is to "extends" and/ or "implements". The other way is to get an instance of the class to achieve the inheritance. That means to make the supposed-super-class be a field member. When you use an instance of the class, actually you get every function available from this class, but you may lose the dynamic features of OOP.

Q.82. What is the difference between final, finally and finalize?

Ans. **Short answer:**

final - declares constant

finally - relates with exception handling

finalize - helps in garbage collection

If asked to give details, explain:

final field, final method, final class

try/finally, try/catch/finally

protected void finalize() in Object class

Q.83. The following statement prints true or false, why?

Ans.

```
byte[] a = { 1, 2, 3 };
byte[] b = (byte[]) a.clone();
System.out.println(a == b);
```

The false will be printed out. Because the two arrays have distinctive memory addresses. Starting in Java 1.2, we can use java.util.Arrays.equals(a, b) to compare whether two arrays have the same contents.

Q.84. What is the output from System.out.println("Hello"+null);?

Ans. Hellonull

Q.85. Is null a keyword?

Ans. The null value is not a keyword.

Q.86. Is sizeof a keyword?

Ans. The sizeof operator is not a keyword.

Q.87. What is clipping?

Ans. Clipping is the process of confining paint operations to a limited area or shape.

Q.88. What is the range of the short type?

Ans. The range of the short type is -(2^15) to 2^15 - 1.

Q.89. What is the range of the char type?

Ans. The range of the char type is 0 to 2^16 - 1.

Q.90. What is the argument type of a program's main() method?

Ans. A program's main() method takes an argument of the String[] type.

Q.91. Which Java operator is right associative?

Ans. The = operator is right associative.

Q.92. What is the difference between the String and StringBuffer classes?

Ans. String objects are constants. StringBuffer objects are not.

Q.93. What is the % operator?

Ans. It is referred to as the modulo or remainder operator. It returns the remainder of dividing the first operand by the second operand.

Q.94. Is &&= a valid Java operator?

Ans. No, it is not.

Q.95. Can there be an abstract class with no abstract methods in it?

Ans. yes.

Q.96. Can an Interface be final?

Ans. yes.

Q.97. Can an Interface have an inner class?

Ans. Yes public interface abc { static int i=0; void dd(); class a1 { a1() { int j; System.out.println("in interfia"); }; public static void main(String a1[]) { System.out.println("in interfia"); } } }

Q.98. Can we define private and protected modifiers for variables in interfaces?

Ans. Yes.

Q.99. Why isn't there operator overloading?

Ans. Because C++ has proven by example that operator overloading makes code almost impossible to maintain. In fact there very nearly wasn't even method overloading in Java, but it was thought that this was too useful for some very basic methods like print(). Note that some of the classes like DataOutputStream have unoverloaded methods like writeInt() and writeByte().

Q.100. What is the main difference between Java platform and other platforms?

Ans. The Java platform differs from most other platforms in that it's a software-only platform that runs on top of other hardware-based platforms.

The Java platform has three elements:

Java programming language

The Java Virtual Machine (Java VM)

The Java Application Programming Interface (Java API)

Q.101. What is the Java Virtual Machine?

Ans. The Java Virtual Machine is a software that can be ported onto various hardware-based platforms.

Q.102. Is Java code slower than native code?

Ans. Not really. As a platform-independent environment, the Java platform can be a bit slower than native code. However, smart compilers, well-tuned interpreters, and just-in-time bytecode compilers can bring performance close to that of native code without threatening portability.

Q.103. Can main() method be overloaded?

Ans. Yes. the main() method is a special method for a program entry. You can overload main() method in any ways. But if you change the signature of the main method, the entry point for the program will be gone.

Q.104. What is JIT?

Ans. JIT stands for Just In Time compiler. It compiles java byte code to native code.

Q.105. What are class members and Instance members?

Ans. Any global members(Variables, methods etc.) which are static are called as Class level members and those which are non-static are called as Instance level members.

Q.106. What is final?

Ans. A final class can't be extended ie., final class may not be subclassed. A final method can't be overridden when its class is inherited. You can't change value of a final variable (is a constant).

Q.107. What if the main method is declared as private?

Ans. The program compiles properly but at runtime it will give "Main method not public." message.

Q.108. What if the static modifier is removed from the signature of the main method?

Ans. Program compiles. But at runtime throws an error "NoSuchMethodError".

Q.109. What if I write static public void instead of public static void?

Ans. Program compiles and runs properly.

Q.110. What if I do not provide the String array as the argument to the method?

Ans. Program compiles but throws a runtime error "NoSuchMethodError".

Q.111. What is the first argument of the String array in main method?

Ans. The String array is empty. It does not have any element. This is unlike C/C++ where the first element by default is the program name.

Q.112. If I do not provide any arguments on the command line, then the String array of Main method will be empty or null?

Ans. It is empty. But not null.

Q.113. What is Overriding?

Ans. When a class defines a method using the same name, return type, and arguments as a method in its superclass, the method in the class overrides the method in the superclass.

When the method is invoked for an object of the class, it is the new definition of the method that is called, and not the method definition from superclass. Methods may be overridden to be more public, not more private.

Q.114. Is Empty .java file a valid source file?

Ans. Yes, an empty .java file is a perfectly valid source file.

Q.115. Can a .java file contain more than one java classes?

Ans. Yes, a .java file contain more than one java classes, provided at the most one of them is a public class.

Q.116. Is String a primitive data type in Java?

Ans. No String is not a primitive data type in Java, even though it is one of the most extensively used object. Strings in Java are instances of String class defined in java.lang package.

Q.117. Is main a keyword in Java?

Ans. No, main is not a keyword in Java.

Q.118. Is next a keyword in Java?

Ans. No, next is not a keyword.

Q.119. Is delete a keyword in Java?

Ans. No, delete is not a keyword in Java. Java does not make use of explicit destructors the way C++ does.

□□□

Previous Year Question Papers

END TERM EXAMINATION

FIFTH SEMESTER (B.TECH.,) DECEMBER-2010

JAVA PROGRAMMING & WEBSITE DESIGN

Time: **3** *Hours* *Total Marks:* **75**

Note: *Attempt five questions. Select one question from each unit including Q.1. which is compulsory.*

1. **(10 × 2.5 = 25)**

(*a*) Briefly explain the compilation process of Java Program, by stating the role of class loader byte code verifier and JSTL.

(*b*) What is the difference between widening and narrowing conversion, given an example of both type.

(*c*) Write any five methods with correct signature of Object class.

(*d*) What is Anonymous Array illustrate it with the help of a program.

(*e*) Give the differences between get and post method of HTTP Request.

(*f*) Write the main drawback of Applets.

(*g*) What are the basic differences between extending the Thread class and implementing the Runnable interface, in order to create a Thread? Which one of these you will prefer to create a Thread.

(*h*) What will happen when we try to execute the following code?

```
class Temp{
Temp(int x){
System.out.println("Constructor"+x);}
Public static void main (String s[]){
Temp t=new Temp(); }} write tour comments for the above code
```

(*i*) Write toyr comments for the below code:

```
class Temp{
    public static vcoid main(String s[]){
    byte b=10;
    byte a=20;
    byte c=a+b;
    System.out.println("C="+c);} }
```

(*j*) Write your Comments for the below code, what will happen when we try to execute the below code

```
class Temp{
    void show(int x, long y){
```

```
        System.out.ptinrln("I Show Method");}
        void show(long x.int y){
        System.out.println("II Show Method"); }
        public static void main(String s[]){
    new Temp().show (10,20);}}
```

Unit-I

2. (*a*) How ASCII code is different from Unicode, convert '*A*' into Unicode format. **6**

(*b*) Create a package mypack which will consist a public class named MYString with a method with signature as **p[ublic String myrew(String ss),** to reverse a String received in argument of the method. Import your package in any class and use this package and method in it, avoid the use of built-in method. Also write the compilation and running syntax of your program. **6.5**

3. (*a*) **(3 × 2 = 6)**

(*i*) What is the difference between equals() and ==

(*ii*) Describe the distributed and dynamic property of java.

(*b*) (*i*) What is the difference between JDK and JRE? How java is platform dependent in case of JRE. **2.5**

(*ii*) Write a program to multiply two matrices using the concept of Anonymos Array. **4**

Unit-II

4. (*a*) What is the difference between checked and unchecked Exception give the example of both type of Exception. How Errors are different from Exception. **6**

(*b*) Write a program to implement the custom Exception. Also explain whether the exceptrion created by usres comes under Checked or Unchecked. **6.5**

5. (*a*) Explain the concept of Deadlock write a program to show the deadlock situation in java. **6.5**

(*b*) Write a program to read from console and write these data into a file, again read that file and print the content onto console.

Unit-III

6. (*a*) Explain the meaning of abstract in AWT. **3.5**

(*b*) Write a program to create the below Structure:

When user will clock on Show button the String Entered in String Text Box should reverse and will be appear in Reverse Text Box. **9**

7. (*a*) Give thne basic difference between Servlet and Applet, write the life cycle of Servlet. **6**

(*b*) Write a program to display the "Hello Servlet" on the webpage using the concept of Servlet, also mayp web.xml, and give the directory Structure of your program/ files. **6.5**

Unit-IV

8. (*a*) Write the differences between Servlet and JSP. **6**

(*b*) Write a program using HTML to create User Registration form, asnd Validate the fields using Java Script use only the (Name (Text Box), Gender (Radio Button), Course (List)) Fields **6.5**

9. (*a*) Write a JSP page to take the name of User if it is equals to IPU then Forward a Welcome Page of it is not equal then forward an Error Page. **6.5**

(*b*) Why we prefer scripting languages in order to check the validity of web pages. **6**

❑❑❑

END TERM EXAMINATION

FIFTH SEMESTER (B.TECH.,) DECEMBER-2009

JAVA PROGRAMMING & WEBSITE DESIGN

Time: **3** *Hours* *Total Marks:* **75**

Note: *Attempt five questions in all Q.No. 1. is compulsory. Attempt one question from each unit.*

1. **(10 × 2.5 = 25)**

(*a*) Explain Garbage collection mechanism in Java?

(*b*) Differentiate throw and throws clause in exception.

(*c*) What is the difference between applet and servlet?

(*d*) Why swings are light weight components?

(*e*) Why do you use jar and manifest file?

(*f*) What is the difference between doGet () and dePost () method.

(*g*) Write a program to show inheritance concept in interface.

(*h*) What is the difference between list and set interfaces.

(*i*) Write a program to show dynamic method dispatch concept.

(*j*) How many types of streams are supported in java. Explain thje topmost classes in different types of streams.

Unit-I

2. (*a*) What are the main features of java programmning language? Give the advantage of java over C++ **6.5**

(*b*) Simulate a calculator using switch statement. **6**

3. (*a*) Write a program to differentiate access modifiers. **6.5**

(*b*) Briefly explain each of the following java terms: **6**

(*i*) String and String buffer.

(*ii*) Abstract and Final

(*iii*) This and Super.

Unit-II

4. (*a*) Write a program to display the content of a gioven file. **6.5**

(*b*) Differentiate between the followings: **6**

(*i*) TCP and UDP

(*ii*) Reader and Writer class

(*iii*) Multithreading and Multiprocessing

5. (*a*) Write a program to show multihreading concept. **6.5**

(*b*) Write a program to show the use of try, catch, throw, throes and finally. **6**

Unit-III

6. (*a*) Write an applety to illustrate the use of the param tag. **6.5**

(*b*) What is the default layout of panel and frame? Describe any 4 layouts. **6**

7. (*a*) Why do you mean by java beans. Define serialization and introspection. Write advantages of making beans over other java programs. **6.5**

(*b*) Explain the Delegation Event Model. Explain the life cycle of a Servlet. **6**

Unit-IV

8. (*a*) Write a short note on electronic mail, internet and intranet services. **6**

(*b*) What are the important things you should consider before designing a good website? **6.5**

9. (*a*) Write HTML form to show radio bottom,s textboxes, checkboxes, tables submit and reset bottom frames. **6.5**

(*b*) Differentiate between the static and dynamic web page. Write the major features of client side scripting. Write coding for an interactive web page. **6**

❑❑❑

BIBLIOGRAPHY

1. Balaguruswamy, E. Objet Oriented Programming in Java, TMH.
2. Gredy Boach, Object Oriented Analysis and Design, Pearson Education.
3. Ram baugh, James, BLAG, EDDY, Object-Oriented Modelling and Design, TMH.
4. Herbert Schildt, The Complete Reference Java 2, TMH.
5. George Ruse, Database Programming with JDBC and Java, O'Reilly.
6. Harold, Java Network programming, O'Reilly.
7. Horstmann CayS., Cornell, Care Java 2 Vol-I and II, Pearsen Education.
8. Downing Troy Brayan, Java RMI: Remake melthod Invocation, IDG Books.
9. Herbers schildt, Joseph O'Neil, JavaBeans programming, TMH.
10. www.Javaworld.com
11. www.sun.java.com

❏❏❏

INDEX

OTHER DETAILS

(i) Who recommended you this book ? (Please tick (✔) in the box near the option relevant to you.

☐ Teacher ☐ Friends ☐ Bookseller

....

....

....

....

(ii) Name of the recommending teacher, his designation and address :

....

....

....

....

(iii) Name and address of the bookseller you purchased the book from :

....

....

....

....

(iv) Name and address of your institution (Please mention the University or Board as the case may be)

....

....

....

....

(v) Your name and complete postal address :

....

....

....

....

(vi) Course you are enrolled in :

....

....

....

....

The best assessment will be awarded each month: The award will be in the form of our publications as decided by the Editorial Board.

Please mail the filled up coupon at your earliest to :

S.K.KATARIA & SONS®
4760-61/23, Ansari Road,
Darya Ganj, New Delhi-110002
☎: 23243489, 23269384
Email : katariabooks@yahoo.com
katson_sanjeev@yahoo.co.in